CAMBRIDGE TEXTS IN THE
HISTORY OF PHILOSOPHY

———

FRANCIS BACON
The New Organon

D0878795

CAMBRIDGE TEXTS IN THE HISTORY OF PHILOSOPHY

Series editors
KARL AMERIKS
Professor of Philosophy at the University of Notre Dame
DESMOND M. CLARKE
Professor of Philosophy at University College Cork

The main objective of Cambridge Texts in the History of Philosophy is to expand the range, variety and quality of texts in the history of philosophy which are available in English. The series includes texts by familiar names (such as Descartes and Kant) and also by less well-known authors. Wherever possible, texts are published in complete and unabridged form, and translations are specially commissioned for the series. Each volume contains a critical introduction together with a guide to further reading and any necessary glossaries and textual apparatus. The volumes are designed for student use at undergraduate and postgraduate level and will be of interest not only to students of philosophy, but also to a wider audience of readers in the history of science, the history of theology and the history of ideas.

For a list of titles published in the series, please see end of book.

FRANCIS BACON

The New Organon

EDITED BY

LISA JARDINE
Queen Mary and Westfield College, University of London

MICHAEL SILVERTHORNE
McGill University

CAMBRIDGE
UNIVERSITY PRESS

PUBLISHED BY THE PRESS SYNDICATE OF THE UNIVERSITY OF CAMBRIDGE
The Pitt Building, Trumpington Street, Cambridge, United Kingdom

CAMBRIDGE UNIVERSITY PRESS
The Edinburgh Building, Cambridge CB2 2RU, UK
http://www.cup.cam.ac.uk
40 West 20th Street, New York, NY 10011-4211, USA
http://www.cup.org
10 Stamford Road, Oakleigh, Melbourne 3166, Australia

© Cambridge University Press 2000

This book is in copyright. Subject to statutory exception and to the provisions of relevant collective
licensing agreements, no reproduction of any part may take place without the written permission of
Cambridge University Press.

First published 2000

Typeset in Ehrhardt 11/13 [CP]

A catalogue record for this book is available from the British Library

Library of Congress cataloguing in publication data

Bacon, Francis, 1561–1626.
[Novum organum. English]
The new organon / Francis Bacon; edited by Lisa Jardine, Michael Silverthorne.
p. cm. – (Cambridge texts in the history of philosophy)
Includes index.
ISBN 0 521 56399 2 (hardback). – ISBN 0 521 56483 2 (paperback)
1. Induction (Logic) Early works to 1800. 2. Science–Methodology Early works to 1800.
I. Jardine, Lisa. II. Silverthorne, Michael. III. Title. IV. Series.
B1168.E5J3713   2000
160 – dc21   99-23266   CIP

ISBN 0 521 56399 2 hardback
ISBN 0 521 56483 2 paperback

Transferred to digital printing 2002

Contents

Preface

My thinking about Francis Bacon's philosophical works has been enormously influenced, and altered in significant ways, by the work I have done over the past five years, leading up to the co-authored biography of Bacon, *Hostage to Fortune: The Troubled Life of Francis Bacon* (1998). That biography was in every sense a collaboration; both the research and the actual writing were conducted as a vigorous partnership between myself and Dr Alan Stewart of Birkbeck College. Accordingly, I acknowledge with deep gratitude here the important part Alan Stewart's research, wisdom and friendship have played in the production of this piece of work.

Lisa Jardine

I would like to thank David Rees, Fellow of Jesus College, Oxford, for his assistance; Julian Martin of the Department of History at the University of Alberta for his help and encouragement in the earlier stages; and Desmond Clarke for his detailed criticism and unfailing courtesy. I owe personal debts of gratitude to Leszek Wysocki at McGill University for the benefit of his expert Latinity, to Katherine Silverthorne for secretarial assistance, and to my wife, Ann, for constant support.

Michael Silverthorne

Introduction

Francis Bacon was born in 1561, the fifth and last surviving son of Sir Nicholas Bacon, Lord Keeper to Queen Elizabeth I, the second surviving child of his second wife. Left a widower in 1552, with six children under twelve to bring up, Nicholas had rapidly married Anne Cooke, one of five highly educated daughters of Edward VI's tutor, Sir Anthony Cooke, celebrated, like their father, for their learning and piety. All made extremely advantageous marriages: Margaret to a prominent goldsmith; Elizabeth to Sir Thomas Hoby and then to the son of the earl of Bedford; Katherine to Sir Henry Killigrew. Most significantly, Mildred became the second wife of William Cecil, later Queen Elizabeth's Principal Secretary of State. Thus Francis was kin to some of the most powerful and influential figures of his time.

This was just as well, since he had to contend, throughout his life, with the fact that his father left him inadequately provided for financially. Sir Nicholas was in the process of making suitable long-term purchases of land for Francis and his elder brother Anthony when he died unexpectedly in 1579. Had that settlement been complete, Bacon later claimed, he would have been able to devote his entire life to study, and his grand plan for an entirely new system of learning might have reached completion in his lifetime. As it was, he was obliged to pursue a civil career, practising law at Gray's Inn in London from the 1580s. Around 1590 Francis and Anthony sought to ballast their financial prospects by entering the service of the earl of Essex as scholarly secretaries and collectors of intelligence.

Given his volatile relationship with the queen, Essex was a poor choice of backer, and in the aftermath of the Essex rebellion (following which

Essex was beheaded), Francis was lucky to survive politically. But then, Bacon was never a good judge of the men in his life. Under the next monarch, James I, he made an equally unwise choice when he threw in his lot with the duke of Buckingham just prior to his disgrace. Bacon's mother complained frequently that he indulged his male servants and turned a blind eye to their petty thieving.

At the time the first edition of *The New Organon* appeared in 1620, as part of a volume publicly presenting in full for the first time the ambitious plan for Bacon's 'Great Instauration', or 'Renewal', of learning, its author was at the high point of his political career. After a series of unsuccessful bids for office under Queen Elizabeth, Bacon's fortunes had slowly improved with the accession of King James in 1603. After 1616, when he became a Privy Councillor and close confidant of the king, his career finally took off. He was made Lord Keeper in 1617 (an office his father had held under Queen Elizabeth), then Lord Chancellor and Baron Verulam in 1618. According to contemporaries, his lifestyle quickly grew in grandeur to match his high office. He kept a vast retinue, dressed his servants and himself with an ostentation verging on the unseemly, and entertained prodigiously and lavishly.[1]

With hindsight, we, of course, know that by the middle of 1621 Bacon had lost everything – impeached, disgraced, briefly imprisoned in the Tower, then banished permanently to the enforced leisure of his country home. But at court and in the intellectual community at large *The New Organon* must have seemed, in October 1620, like the culminating achievement of one of the brightest stars in England's political firmament.

The work's ambitiousness matched that of its author; nor did Bacon separate this bid for worldwide intellectual recognition from his more parochial aspirations at the English court. It would, he hoped, further cement the king's personal commitment to him. In a private letter to James accompanying the presentation copy, Bacon flattered him with the suggestion that he was the person at whom the entire 'Great Instauration' was directed:

The work, in what colours soever it may be set forth, is no more than a new logic, teaching to invent and judge by induction (as finding syllogism incompetent for sciences of nature), and thereby to make philosophy and sciences both more true and more active. This, tending to enlarge the bounds of Reason and endow man's

[1] For Bacon's biography see, most recently, L. Jardine and A. Stewart, *Hostage to Fortune: The Troubled Life of Francis Bacon* (London, Gollancz, 1998).

estate with new value, was no improper oblation to your Majesty, who, of men, is the greatest master of reason, and author of beneficence.[2]

A history of experiments

A 'Great Renewal' of learning written by the Lord Chancellor was guaranteed a serious reception as a work of philosophy. In December 1620, just two months after its official publication date, the English diplomat Henry Wotton, on a mission to Vienna, wrote to Bacon to acknowledge safe receipt of three copies of *The New Organon*.

He was not yet in a position to comment on the philosophical work, Wotton apologised, 'having yet read only the first Book thereof, and a few Aphorismes of the second'. For the time being, therefore, he would instead make a modest practical contribution to the Lord Chancellor's grand scientific project, for which *The New Organon* was to provide the methodological infrastructure. Wotton had apparently agreed to send Bacon reports of interesting scientific matters he encountered in the course of his embassy: 'I owe your Lordship even by promise (which you are pleased to remember, thereby doubly binding me) some trouble this way: I mean by the commerce of *Philosophical* experiments, which surely, of all other, is the most ingenious Traffick.'

Wotton went on to tell Bacon that he had just returned from a visit to the home of the famous astronomer Johannes Kepler. He had, indeed, decided to present Kepler with one of the copies of *The New Organon* sent by Bacon, 'that he may see that we have some of our own that can honour our King, as well as he hath done with his *Harmonice Mundi*'. In Kepler's study Wotton had seen a most remarkable piece of scientific apparatus in action:

He hath a little black tent, exactly close and dark, save at one hole, about an inch and an half in the Diameter, to which he applies a long perspective-tube, with the convex glasse fitted to the said hole, and the concave taken out at the other end, through which the visible radiations of all the objects without are intromitted falling upon a paper, which is accommodated to receive them. And so he traceth them with his Pen in their natural appearance, turning his little Tent round by degrees till he hath designed the whole aspect of the field.[3]

[2] J. Spedding (ed.), *Letters and Life* (7 vols., London, Longman, Green, Longman and Roberts, 1861–74), 7, 119–20.

[3] L. Pearsall Smith (ed.), *The Life and Letters of Sir Henry Wotton* (2 vols., Oxford, Clarendon Press, 1907), 2, 206. See also S. Alpers, *The Art of Describing: Dutch Art in the Seventeenth Century* (Chicago, University of Chicago Press, 1983), 49–51.

Here, clearly, was a device with a whole range of possible applications. Kepler's 'little black tent' was a refined version of a favourite piece of seventeenth-century European new technology, the *camera obscura*. It would, Wotton pointed out, be a particularly useful technical tool for covertly drawing accurate maps and harbour plans.

Wotton's letter (which Bacon's nineteenth-century editor Spedding omitted from his definitive edition)[4] shows how Bacon's political status at home allowed him unusually direct access to emerging seventeenth-century science on the mainland of Europe. It belongs alongside two letters sent by Bacon's friend Toby Matthew from Italy, the first written in 1616 (also omitted from the *Works* by Spedding) and reporting Galileo's support for Copernican astronomy, the second informing Bacon that Galileo had produced a written response to Bacon's own paper on the ebb and flow of tides. Together they confirm that Bacon did not simply comment on, but took an active part in, what stands today as cutting-edge European scientific thought of his day.

Although Wotton had barely begun reading the second book of *The New Organon* when he wrote to Bacon, he was correct in anticipating that the *camera obscura* was not mentioned among the examples of new technology in the 'privileged instances' section of the book. Bacon did, however, there discuss a related piece of lens technology, the microscope, , in some detail. Among the 'privileged instances' which specially advance the investigation of nature, Bacon lists 'instances that open doors or gates' – instances which 'assist direct actions of the sense'. The microscope falls within this category of valuable aids to the senses:

Apart from spectacles and such things, whose function is simply to correct and alleviate the weakness of impaired vision, and so provide no new information, [such] an instance ... are microscopes, lately invented, which (by remarkably increasing the size of the specimens) reveal the hidden, invisible small parts of bodies, and their latent structure and motions. By their means the exact shape and features of the body in the flea, the fly and worms are viewed, as well as colours and motions not previously visible, to our great amazement. (II.39)

Bacon's information on the moving parts of tiny organisms made visible by the microscope came from closer to hand than the written reports he

---

4 Spedding keeps Bacon's 'life' (of which his letters are obviously a crucial part) separate from his 'works' (his published science and philosophy). He therefore consistently excludes letters which refer directly to Bacon's scientific practice. See L. Jardine and A. Stewart, 'Judge Him According to His Works: James Spedding's Textual Defence of Francis Bacon' in N. Jardine and M. Frasca-Spada (eds.), *History of the Sciences/History of the Book* (Cambridge, Cambridge University Press, in press).

received of Continental scientific activity. William Harvey (of circulation-of-the-blood fame) recorded in his landmark work *On the Motion of the Heart and Blood in Animals*, published originally in Latin as *De motu cordis*, two years after Bacon's death, in 1628:

In bees, flies, hornets, and the like, we can perceive something pulsating with the help of a lens; in *pediculi* [little lice], also, the same thing may be seen, and as the body is transparent, the passage of the food through the intestines, like a black spot or stain, may be perceived by the aid of the same magnifying lens.[5]

For Harvey the value of such observations was as 'ocular' confirmation that the heart acts as a pneumatic pump, driving the blood in a perpetual circuit around the body of an animal, whatever its size. The anatomical investigations which clinched his revolutionary discovery were, of course, conducted on much larger-scale organisms. As Physician Extraordinary to James I (and later to his son Charles), Harvey had access to a wide and exotic range of animals on whose living and dead bodies he could experiment. By 1616, when he first outlined the theory of blood circulation in the Lumleian Lecture to the Royal College of Physicians, he had dissected not only human cadavers but innumerable royal deer and an ostrich from the king's menagerie; he had also conducted simple vivisectional experiments on domestic animals and excised the hearts of live vipers.[6]

Bacon was at one time one of Harvey's aristocratic patients (in 1619 both Bacon and the king suffered extended bouts of the stone).[7] According to John Aubrey, Harvey recalled that Bacon had the cold hazel eyes of the vipers on his dissection table. It was also Harvey who observed to Aubrey of Bacon that 'he writes philosophy like a Lord Chancellor' – which comment, Aubrey adds, was meant 'in derision'. In *The New Atlantis*, it is the Royal Physician's anatomical dissections which Bacon has in mind when he describes how his ideal scientists in the imaginary land of Bensalem kept 'inclosures of all sorts of beasts and birds, which [they] use not only for view or rareness, but likewise for dissections and trials; that thereby [they] may take light what may be wrought upon the body of man'.[8] In *The New Organon*, it is Harvey to

---

[5] R. Willis (trans.) and A. C. Guyton (ed.), *The Works of William Harvey* (Philadelphia, University of Pennsylvania Press, 1989), 76.

[6] R. G. Frank, Jr, *Harvey and the Oxford Physiologists: A Study of Scientific Ideas* (Berkeley and Los Angeles, University of California Press, 1980), 1.

[7] See Jardine and Stewart, *Hostage to Fortune*, 425–7.

[8] J. Spedding, R. L. Ellis and D. D. Heath (eds.), *Works* (7 vols., London, Longman et al., 1857–9), 3, 159.

whom Bacon also alludes when he remarks, in the context of observing the natural processes of generation and growth in animals:

It would be inhuman to make such investigations of well-formed animals ready for birth, by cutting the foetuses out of the womb, except for accidental abortions and in hunting, and so on. (II.41)

To understand *The New Organon* in the spirit in which it was written, we need to be clear that it is driven by a strong commitment to new technical scientific instruments and the increasing variety of experiments on nature they made possible. *The New Organon* belongs to an early-seventeenth-century English intellectual milieu which included William Gilbert's work with magnetism and Harvey's on the circulation of the blood, which took into account technological innovation on the Continent, and which looked forward to Robert Boyle's and Robert Hooke's experiments with air-pumps in the 1650s. It is an extraordinary attempt to give formal shape to a rapidly emerging (but hitherto largely problem-driven and *ad hoc*) new experimentally based science.

The new scientific method

Bacon's *Novum Organum*, or *The New Organon*, takes its title from Aristotle's work on logic, the 'Organon' or 'Instrument for Rational Thinking'. Bacon argued vigorously in his *Advancement of Learning* that Aristotle's logic was entirely unsuitable for the pursuit of knowledge in the 'modern' age. Accordingly, *The New Organon* propounds a system of reasoning to supersede Aristotle's, suitable for the pursuit of knowledge in the age of science. Where Aristotle's inferential system based on syllogisms could reliably derive conclusions which were logically consistent with an argument's premises, Bacon's system was designed to investigate the fundamental premises themselves. Aristotle's logic proposed certainty, based on incontrovertible premises accepted unquestioningly as true; Bacon proposed an inductive inference, based upon a return to the raw evidence of the natural world. From painstakingly collected assemblages of data ('natural histories') the scientific investigator would use *The New Organon* to nudge his way gradually towards higher levels of probability.[9]

[9] Bacon reminds his reader a number of times in *The New Organon* that his method bears a resemblance to that adopted by the ancients' 'weak scepticism'. On weak scepticism as understood in the Renaissance see L. Jardine, 'Lorenzo Valla: Academic Skepticism and the New Humanist Dialectic' in M. Burnyeat (ed.), *The Skeptical Tradition* (Berkeley and Los Angeles, University of California Press, 1983), ch. 10, 253–86.

The New Organon was the second part of the six-part programme of scientific inquiry assembled under the title *The Great Instauration*, or 'Great Renewal', of learning. As Bacon explains in the published introduction to the *The New Organon* volume, the first part was to consist of 'the divisions of the sciences', comprising 'a Summary or general description of the science or learning which the human race currently possesses' ('Plan of the Work'). Since this part of the enterprise remained incomplete in 1620 (as indeed it did at Bacon's death), the title page to *The New Organon* explains that some account of them will be found in the Second Book of *The Advancement of Learning*.

Beyond the second part of *The Great Instauration* as presented in *The New Organon*, all further parts remained, on Bacon's own admission, woefully incomplete. The third part was to be a comprehensive compilation of 'the *Phenomena of the Universe*, that is, every kind of experience, and the sort of natural history which can establish the foundations of philosophy' ('Plan'). Such a comprehensive collection of natural-historical data would include not simply all natural phenomena ('a history of the bodies of heaven and the sky, of land and sea, of minerals, plants and animals'), but also compilations of the material in all existing academic disciplines, and histories of crafts, trades and other examples of nature 'pressured and moulded' (ibid.). In several places in his works and letters, Bacon stresses that this project is a massive collaborative one, requiring the financial and organisational backing of a 'King or Pope'.[10]

Using the material from this databank, the fourth part was to be a collection of disparate and preliminary worked examples of the 'method' of *The New Organon* in action. This is the vaguest of Bacon's sections, and our understanding of it is based on a single paragraph in the 'Plan', together with some equally sketchy remarks in a letter written shortly before his death. There Bacon describes Part Four as 'an intellectual machinery' comprising 'axioms and higher level observations' of a kind already to be found in his own fragmentary Histories, but 'better adjusted to the rules of the inductive method'.[11] Presumably it would be a series of 'systems', of limited explanatory force – like Gilbert's based on the principle of magnetism or Bacon's own based on his principle of motion. Ultimately such partial explanations would all be subsumed under the totalising explanatory system of the sixth part, where 'the philosophy which is

[10] 'Epistola ad Fulgentium' in Spedding, *Letters and Life*, 7, 531.
[11] Spedding, *Letters and Life*, 7, 532.

derived and formed from the kind of correct, pure, strict inquiry ... already framed' is brought to completion ('Plan'). This would finally allow mankind unlimited power to control the natural world not by coercion but by complete understanding:

> For man is nature's agent and interpreter; he does and understands only as much as he has observed of the order of Nature in work or by inference; he does not know and cannot do more. No strength exists that can interrupt or break the chain of causes; and nature is conquered only by obedience. (Ibid.)

Completing the sixth part of *The Great Instauration* himself is, Bacon confesses, a thing both 'beyond our ability and beyond our expectation'. For the time being, the fifth part will consist of provisional discoveries, of the kind made by Bacon himself. These would ultimately be made obsolete by the grand design, but in the meantime would provide encouragement for those looking for tangible results.

Bacon's metaphor of such results being like interest payable upon a capital sum invested, which keep the investor going until the capital itself is redeemed, indicates the strenuously pragmatic thinking behind his philosophical undertaking. The kinds of investors he seeks for his *Great Instauration* need to know they can expect to make a rapid profit. In the long term they may be prepared to indulge the Lord Chancellor in his pursuit of a single overarching system to explain the entire natural world. In the short term, they will need to see immediate pay-off in the form of enhanced procedures in traditional trade and manufacture. It was certainly this aspect of the Baconian project which Charles II, returning in 1660 to a financially weakened England after the Commonwealth period, found particularly attractive when he agreed to give his name and political backing to the Bacon-inspired Royal Society (the 'Royal Society of London, for Improving of Natural Knowledge', to give it its full title).[12]

Observation and experiment

In spite of its reputation as a single-mindedly theoretical work of scientific epistemology, over half of Bacon's *New Organon* is taken up with examples

---

[12] For the early aspirations of the Royal Society see M. Hunter, 'The Significance of the Royal Society', *Science and Society in Restoration England* (Cambridge, Cambridge University Press, 1981), 32–58. On Baconianism and the Restoration see also C. Webster, *The Great Instauration: Science, Medicine and Reform 1626-1660* (London, Duckworth, 1975).

from applied science which can be traced to contemporary experimental work in a broad range of emerging fields.[13]

What distinguishes the new Baconian view of science (as presented most clearly in *The New Organon*) from that of his predecessors is, indeed, his clear commitment to the role of observation and experiment as a pre-requisite for the construction of scientific theory itself. Earlier scientists (and scientific near-contemporaries elsewhere in Europe) had thought of observation and experiment as demonstrating a conclusion anticipated by systematic deductive reasoning, or as determining a detail or filling in a gap, as required to extend an existing theory. Thus, for instance, Robert Boyle (a keen follower of Bacon) was quick to point out that Blaise Pascal's 'experiments' in hydrostatics, adduced in support of his theoretical principles, are clearly impossible-to-perform 'thought experiments' whose proposed outcomes are calculated to confirm an already decided theory. Bacon, by contrast, regarded observation and experiment – particularly experiments designed to test how nature would behave under previously unobserved circumstances – as the very foundation of science and its generalised methodology. He expected that the process itself of organising the mass of data collected into natural and experimental histories would lead to an entirely new and largely unforeseen scientific theory.[14]

Among such groundbreaking experiments included in *The New Organon* are a number which Bacon had clearly carried out himself, mostly experiments in chemistry and mechanics (he may, however, like Boyle, have had laboratory assistants who actually performed the experiments, while he himself simply observed, as befitted a gentleman). For example, in Bacon's discussion of specific gravity, under 'privileged instances' we find the following description of an experiment using equipment familiar to the seventeenth-century alchemist or chemist:

We took a small glass bottle, which could hold perhaps one ounce (we used a small vessel so that the consequent evaporation could be achieved with less heat). We

---

[13] For the kind of early scientific tradition to which Bacon's practical experimental experience belongs see Frank, *Harvey and the Oxford Physiologists*, and S. Shapin and S. Schaffer, *Leviathan and the Air-Pump: Hobbes, Boyle, and the Experimental Life* (Princeton, NJ, Princeton University Press, 1985). It has recently been pointed out that Bacon was relatively cut off from a specifically Continental tradition in published natural history. See P. Findlen, 'Francis Bacon and the Reform of Natural History in the Seventeenth Century' in D. Kelley (ed.), *History and the Disciplines: The Reclassification of Knowledge in Early Modern Europe* (Rochester, NY, University of Rochester Press, 1997), 239–60.

[14] See T. S. Kuhn, 'Mathematical versus Experimental Traditions in the Development of Physical Sciences', *The Essential Tension: Selected Studies in Scientific Tradition and Change* (Chicago, University of Chicago Press, 1977), 31–65.

filled this bottle with spirit of wine almost to the brim; selecting spirit of wine because we observed by means of an earlier table that it is the rarest of the tangible bodies (which are continuous, not porous), and contains the least matter for its dimensions. Then we accurately noted the weight of the liquid with the bottle itself. Next we took a bladder which would hold about two pints. We expelled all the air from it, so far as possible, to the point that the sides of the bladder were touching each other; we had also previously smeared the bladder with grease, rubbing it gently in so that it would be more effectively closed, its porosity, if there was any, being sealed by the oil. We tied this bladder tightly around the mouth of the bottle, with its mouth inside the mouth of the bladder, lightly waxing the thread so that it would stick better and bind more tightly. Finally we placed the bottle above burning coals in a brazier.

Bacon goes on to describe the carefully observed and quantified outcome of the experiment:

Very soon a steam or breath of spirit of wine, expanded by heat and turned into gaseous form, gradually inflated the bladder, and stretched the whole thing in every direction like a sail. As soon as this happened, we removed the glass from the fire, and placed it on a rug so that it would not be cracked by the cold; we also immediately made a hole in the top of the bladder, so that when the heat ceased, the steam would not return to liquid form and run down and spoil the measurement. Then we lifted up the bladder itself and again took the weight of the spirit of wine which remained. Then we calculated how much had been used up as steam or gas; and making a comparison as to how much place or space that substance had filled in the bottle when it was spirit of wine, and then how much space it filled after it had become gas in the bladder, we calculated the ratio; and it was absolutely clear that the substance thus converted and changed had achieved a hundredfold expansion over its previous state. (II.40 (3, 4))

This is a perfectly respectable experiment in expansion of gases, of the kind Robert Boyle would be performing in Oxford by the 1650s. It involves calibration and quantification: spirit of wine, or brandy, is chosen because it vaporises at low temperatures; the quantity of brandy vaporised is calculated by weighing the container before and after heating; the volume to which the vaporised brandy expanded is calculated by calculating the volume of the bladder. The finding was that the vaporised brandy 'had achieved a hundredfold expansion over its previous state'.

Bacon lets his reader know clearly, by the formal locutions he uses, when it is he who has conducted experiments and when he has merely 'heard tell' or read about them at second hand. In another of his own experiments, described under 'privileged instances' of 'range or furthest limit', he tested 'how much compression or expansion bodies easily and freely allow (in accordance with their natures), and at what point they begin to resist, so that at the extreme they bear it No Further [*ne plus ultra*]':

as when an inflated bladder is compressed, it tolerates some compression of the air, but after a point it can bear it no longer, and the bladder bursts.

We tested this more precisely with a subtle experiment. We took a small metal bell, quite thin and light, like a saltcellar, and sank it in a basin of water, so that it took with it to the bottom of the basin the air held in its cavity. We had first placed a little ball on the bottom of the basin on which to set the bell. The result was that if the ball was quite small (in relation to the cavity) the air retreated into a smaller area, and was simply compressed and not expelled. But if it was too large for the air to give way freely, then the air could not tolerate the greater pressure but partially lifted the little bell and came up in bubbles. (II.45)

This kind of experiment with bells is related to contemporary practical experiments with diving-bells. Sure enough, under 'multi-purpose instances' Bacon refers directly to the technology of diving-bells for salvage operations:

If the situation requires bodies to be submerged in a depth of water, a river perhaps or the sea, but not to have contact with the water, and not to be shut up in sealed vessels but to be just surrounded by air, very useful is the vessel which is some-times used to work under the water on sunken ships, which enables divers to stay under water longer and to take breaths in turn from time to time. It was like this. A concave metal barrel [bell] was constructed, and was let down evenly into the water, its mouth parallel to the surface; in this way it carried all the air it contained with it to the bottom of the sea. It stood on three feet (like a tripod) which were a little shorter than a man, so that when a diver ran out of breath, he could put his head into the hollow of the jar, take a breath, and then continue with his work. We have heard that a device has just been invented like a small ship or boat, which can carry men under water for a certain distance. Under the kind of jar we mentioned above, certain bodies could easily be suspended; that is why we adduced this experiment. (II.50 (I))

At the end of this passage we have evidence that Bacon was aware of a celebrated English demonstration of Cornelis Drebbel's 'submarine' ('we have heard that ...') and that he anticipated the kinds of experiments in an evacuated chamber carried out forty years later using Boyle's air-pump.

The coherence of Bacon's new scientific method, upon which his 'Great Renewal' of learning rests, depends upon a close relationship between experimental practice and methodical processing of results. This was clearly understood by the early Royal Society in London, for whom the great Verulam (Sir Francis Bacon) was figurehead, patron saint and 'Father of Modern Science'. Early Restoration scientists like Royal Society Curator of Experiments Robert Hooke took *The Great Instauration* entirely seriously and modelled their own programmes for an experimentally based science directly on Bacon's writings on methodology.[15]

By the end of the nineteenth century, however, attention had become focused almost exclusively upon the formal validity of the inductive method of *The New Organon*, and its adequacy as a substitute for the logico-deductive epistemology which it supposedly superseded.[16] In most modern accounts of Baconian method, the groundbreaking originality of Bacon's direct engagement with contemporary applied science and technology, leading to his attempt to devise an epistemology which reflected the intimate relationship in science between ideas and practice, has been lost from sight.[17]

Baconian induction

The New Organon is laid out in two books of 'aphorisms'. These are relatively concise, unembellished statements designed to indicate that this is 'work in progress', susceptible of improvement and refinement, rather than a finished and coherent 'Organon'. The aphoristic 'method of delivery' ensures that the reader understands that the work is incomplete, and also reveals clearly the competence of the author. As Bacon explained three years later, in the expanded Latin version of the *Advancement of Learning*:

Delivery by aphorisms ... tries the writer, whether he be light and superficial in his knowledge or solid. For aphorisms, not to be ridiculous, must be made out of the pith and heart of sciences ... A man will not be equal to writing in aphorisms, nor indeed will he think of doing so, unless he feels that he is amply and solidly furnished for the work.[18]

---

[15] On Hooke and Bacon's legacy see L. Jardine, '*Experientia literata* or *Novum Organum?* The Dilemma of Bacon's Scientific Method' in W. Sessions (ed.), *Francis Bacon's Legacy of Texts* (New York, AMS Press, 1990), 47–68.

[16] This position may be represented by Hall's classic treatment in A. R. Hall, *The Revolution in Science 1500–1750* (London, Longman, 1983) (previously published as *The Scientific Revolution* (1954)). See esp. ch. 7 ('New Systems of Scientific Thought in the Seventeenth Century').

[17] For later philosophers' attitudes to Bacon's inductive method see A. Pérez-Ramos, *Francis Bacon's Idea of Science and the Maker's Knowledge Tradition* (Oxford, Clarendon Press, 1988), chs. 17 and 18.

[18] Spedding et al., *Works*, 4, 450–1.

Aphorisms have the further advantage that they encourage collaborative pursuit of knowledge and 'invite others to contribute and add something in their turn'.

Book One of *The New Organon* clears away the intellectual debris of existing assumptions which distort the perceptions and cloud the judgement of the would-be philosopher. Bacon urges readers not to place their trust in existing authorities, nor to rush to fashionable new systems of knowledge. 'A new beginning [to learning] has to be made from the lowest foundations, unless one is content to go round in circles for ever' (1.31).

To this end, Bacon first discards the cornerstone of traditional logic, the syllogism, because anyone using it can only arrive at conclusions consistent with existing, given premises. These premises themselves – the assertions on which the whole process of reasoning is based – must be taken on trust as true and incontrovertible. The syllogism 'is not applied to the principles of the sciences, and is applied in vain to the middle axioms, since it is by no means equal to the subtlety of nature' (1.13). Thus the entire current system of reasoning, in Bacon's view, fails. Instead Bacon gives notice that his own logic will be an induction or gradual ascent from sense data to generalisations, though not the 'ordinary induction' of the logic handbooks.[19]

A significant portion of Book One of *The New Organon* is taken up with discussion of what Bacon names the 'Idols', or 'Illusions' – impediments of various kinds which interfere with the processes of clear human reasoning. These so-called Idols are of four kinds: Idols of the Tribe, Idols of the Cave, Idols of the Marketplace and Idols of the Theatre.

The Idols of the Tribe are errors in perception itself, caused by the limitations of the human senses which give access to the data of nature. The Idols of the Cave, by contrast, are errors introduced by each individual's personal prejudices and attachment to particular styles or modes of explanation – as in his fellow-courtier (and personal physician to Elizabeth I) William Gilbert's trying to account for all natural phenomena in terms of magnetism.[20]

[19] On induction in the logic manuals see L. Jardine, 'Humanistic Logic' in C. B. Schmitt, E. Kessler and Q. R. D. Skinner (eds.), *The Cambridge History of Renaissance Philosophy* (Cambridge, Cambridge University Press, 1988), 173-98. On Bacon's rejection of the syllogism see L. Jardine, *Francis Bacon: Discovery and the Art of Discourse* (Cambridge, Cambridge University Press, 1974), 84-7.

[20] William Gilbert (*c.* 1540–1603), personal physician to Elizabeth I, is known for his early studies on electricity and magnetism. He received his M.D. in 1569 from Cambridge University, where he had been made a Fellow of St John's College in 1561; in 1600 he was elected President of the Royal College of Physicians. In the same year he published *De magnete*, in which he propounded the theory that the earth was a giant lodestone with north and south magnetic poles.

The Idols of the Marketplace arise directly from shared use of language and from commerce between people. At the most basic level, the ascription of names to things, in ordinary language usage, fails to discriminate properly between distinctive phenomena, or names abstract entities 'vaguely', so as to give rise to false beliefs about them.

Finally, Idols of the Theatre are the misleading consequences for human knowledge of the systems of philosophy and rules of demonstration (reliable proof) currently in place. These Bacon insists are 'so many plays produced and performed which have created false and fictitious worlds' (1.44). It is here that Bacon expounds his antipathy towards existing philosophies, most notably the Aristotelian system which exercised a virtual stranglehold on contemporary thought, but extending also to Platonism and the methods of philosophical doubt of the Sceptics.[21]

The final aphorisms of Book One deal with the characteristics of a scientific method which will be adequate to handle the proliferation of innovation in practical science, and to process the 'stock and material of natural history and experience' (1.101). It will be based on improved natural histories (to be described in the third part of *The Great Instauration*), and it will employ an entirely new form of induction (the subject of Book Two of *The New Organon*). Used together, these will revolutionise learning and finally give mankind that power over nature of which early scientists such as the alchemists could only dream.

Book One of *The New Organon* contains some shrewd insights, particularly into the shortcomings of burgeoning science as currently practised. Bacon points out, for instance, that the alchemist, attempting to transmute base metal into gold, commits himself to performing a single chain of experimental operations according to secret instructions. When this does not succeed, the practitioner does not discard the recipe; rather 'He accuses himself of not properly understanding the words of the art or of the authors ... or of making a slip in the weights or timing of his procedure, and so proceeds to repeat the experiment indefinitely' (1.85).[22] Not only is this a perceptive description of seventeenth-century chemical procedures, it also suggests that Bacon had himself observed contemporary proto-chemists at work in the laboratory.[23]

---

[21] For Bacon's disdainful attitude to ancient philosophies see *Advancement of Learning*, Book One.

[22] For the accuracy of this account of alchemical practice, compare L. M. Principe, *The Aspiring Adept: Robert Boyle and his Alchemical Quest* (Princeton, NJ, Princeton University Press, 1998).

[23] Bacon's observation of the alchemist's practice anticipates Thomas Kuhn's recognition that the scientist does not generally discard a 'paradigm' or model for scientific practice merely because he is unable to replicate predicted outcomes. See T. S. Kuhn, *The Structure of Scientific Revolutions* (Chicago, University of Chicago, 1962).

The goal of Bacon's inductive method is the 'forms', or essential definitions, of what he terms 'simple natures' – the fundamental building blocks out of which all compound bodies are, in his view, constructed. The procedure for finding the form of a simple nature is clearly set out at the beginning of Book Two of *The New Organon*. The investigator starts by collecting into a History all available occurrences of the nature selected (Bacon chooses the simple nature 'heat' for his worked example). From among these he selects those which provide as clear as possible a picture of the nature's production. These are organised, tabulated and collated and any gaps filled in with examples drawn from specially designed experiments. Together they form a 'table of presence'.

A second table is now drawn up, in which the instances of presence of the selected nature are matched as closely as possible by ones from which it is absent (if the rays of the sun are an instance of presence of heat, the rays of the moon are an instance of absence). These two tables are supplemented by one which lists instances where the increase or decrease of the nature is accompanied by increase or decrease of other properties present, indicating that these may be essential concomitants of the nature under investigation. Bacon's 'induction' is carried out by eliminating extraneous and redundant material between the three tables, to yield an essential physical description of the simple nature – its 'form'. This simple elimination is the unique legitimate use of formal inference in the entire interpretation of nature.[24]

All further steps in the process of refining the 'simple natures' and their essential 'forms' depend directly on the observed outcomes of carefully classified experiments. There will, Bacon makes clear, inevitably be such further stages, until at some yet-to-be-determined date in the distant future a single overarching explanatory theory is arrived at (by years of assiduous practice and experiment).

In the meantime, of the fifty-two aphorisms in Book Two of *The New Organon* only the first twenty are taken up with the inductive method. The remaining aphorisms compile a collection of 'privileged instances' under various categories. These are types of experimental set-up which provide particularly powerful kinds of tool for investigating nature.

'Privileged instances'

At this stage in the presentation of his 'new instrument' for the interpretation of nature, Bacon asks us, with characteristic intellectual candour, to

[24] For a full account of Baconian method see Jardine, *Francis Bacon*, esp. ch. 6.

adopt a vigorously pragmatic approach to achieving a 'true and complete induction'. The process of assembling tables of presence, absence and variation for instances of any given nature and eliminating between them will yield only a 'first harvest', or preliminary interpretation. What follows is a series of types of 'support' for the understanding. It is these 'privileged instances' which will extend and refine the preliminary findings into a valid 'form'. There are a number of types of instance of occurrence of any given simple nature which reveal aspects of that nature with striking clarity. Such instances allow the investigator to move decisively and particularly swiftly towards identifying the fundamental characteristics which make up the 'form' of the nature.

According to Bacon, what makes the twenty-seven types of 'privileged instances' of such significance is that they allow the investigator to guide the later stages of induction by using 'the nature of things' themselves:

Our logic instructs the understanding and trains it, not (as common logic does) to grope and clutch at abstracts with feeble mental tendrils, but to dissect nature truly, and to discover the powers and actions of bodies and their laws limned in matter. Hence this science takes its origin not only from the nature of the mind but from the nature of things. (II.52)

A preliminary attempt at inductive solution begins the process. Privileged instances direct and steer the investigation further in the right direction.

Those who have looked to Bacon for a genuinely new logic of scientific inquiry have generally ignored the 'privileged instances', since their procedures, and the guidance they yield towards forms of simple natures, are, on Bacon's own admission, observation- and experiment-led and *ad hoc*. Besides, they reach their supplementary conclusions by conventional deduction rather than by any kind of induction. Their relationship to the preliminary 'first harvest' most closely resembles the way individual cases at law are used to refine legal precept in English case law: a broadly applicable rule is agreed upon, and its impact is refined by using the detail of successive cases for clarification.[25]

In the light of developments in natural science later in the seventeenth century and the so-called scientific revolution, however, the 'privileged instances' of *The New Organon* deserve particular attention for the

---

[25] On the link between Bacon's inductive method for science and the procedures of the contemporary law-courts in which Bacon practised throughout his professional life see D. Coquillette, *Francis Bacon* (Stanford, CA, Stanford University Press, 1992).

evidence they provide of Bacon's being abreast of hotly debated issues in contemporary science, and for his clear sense that they provide proper guidance in a generalised methodology for the sciences. Under 'instances of alliance' (ones which determine whether phenomena which tend to occur together do so as cause and effect or accidentally) he discusses Gilbert's contention that all dense and solid bodies are magnets which move towards the earth (itself a magnet) by attraction as long as they are within the 'circle of its own power' (II.35). The giant waterspouts observed en route to the East and West Indies might suggest that at sufficient height above the earth water escapes from its attractive force. Later, under 'instances of range or furthest limit', Bacon returns to the same topic. Iron within a certain distance of a magnet is drawn towards it, but not beyond. If the earth is a giant magnet, its powers will extend to a great distance; similarly for the moon's attractive force on the waters of the sea, which causes high and low tides:

But whether the distance at which they work is great or small, all these things certainly work at distances which are fixed and known to nature, so that there is a kind of *No Further* which is in proportion to the mass or quantity of the bodies; or to the vigour or weakness of their powers; or to the assistance or resistance of the surrounding medium; all of which should come into the calculation [be calculated] and be noted. (II.45)

Scattered through the 'privileged instances', observations like these reveal Bacon's clear sense of what it would take methodologically to confirm or disprove a proposed theory.

Certain of Bacon's categories of 'privileged instance' stand out as fore-runners of later standard scientific procedures. His 'crucial instances' (the fourteenth of his types) list a selection of carefully designed experimental set-ups on the basis of which the scientist can decide between alternative views concerning the phenomena under investigation. These Bacon also characterises as 'decisive instances', 'instances of verdicts' and 'commanding instances':

Sometimes in the search for a nature the intellect is poised in equilibrium and cannot decide to which of two or (occasionally) more natures it should attribute or assign the cause of the nature under investigation ... in these circumstances crucial instances reveal that the fellowship of one of the natures with the nature under investigation is constant and indissoluble, while that of the other is fitful and occasional. This ends the search as the former nature is taken as the cause and the other dismissed and rejected. (II.36)

It is here that Bacon discusses the nature of the ebb and flow of tides at length, incorporating his exchanges with Galileo and proposing several decisive tests for competing contemporary conjectures.

The systematic guidance afforded by procedures like those itemised under his 'privileged instances' belongs to what Bacon terms *experientia literata* – 'experience made literate'. *Experientia literata* allows the investigator to organise the material gathered together in a given History so as to extend the History itself (to yield a more precise definition of the 'form' under investigation). *Experientia literata* also throws up practical applications and scientific innovations (for trade and commercial benefit) in advance. In both cases, the process of reasoning which leads from particular experiment or instance to generalisation is, according to Bacon, guided by connections intrinsic to nature itself rather than connections imagined by the human mind. *Experientia literata* thus binds together the stages in ratiocination laid down for the 'new instrument' and the material world to which it is to be applied.

Benefits derived during the process of systematic investigation are, however, Bacon stresses, still only provisional. In the end, *experientia literata* is merely a stage on the way to 'forms':

For although I do not deny that when all the experiments of all the arts shall have been collected and digested, and brought within one man's knowledge and judgment, the mere transferring of the experiments of one art to others may lead, by means of that experience which I term 'literate', to the discovery of many new things of service to the life and state of man, yet it is no great matter that can be hoped from that; but from the new light of axioms [generalisations], which having been educed from those particulars by a certain method and rule, shall in their turn point out the way again to new particulars, greater things may be looked for.[26]

We look in vain for clear guidance in *The New Organon* as to how the methodological derivation of 'forms' and the axioms and rules which connect them will eventually be brought to completion. Towards its close, Book Two becomes a checklist of yet-to-be-investigated possibilities for sharpening the outcome of Bacon's inductive method – a collection of Lord Chancellor's jottings towards a future enlarged project. Like the 'Plan' which precedes *The New Organon* in the first printed edition, and the inventory of 'Natural and Experimental Histories' which follows it, the

[26] *De augmentis scientiarum*; Spedding et al., *Works*, 4, 96.

philosophical method of the 'new instrument' is itself (as the presentation in aphoristic form is supposed to indicate) unfinished.

In many respects Baconian methodology has little relevance to the kinds of debate around epistemology which preoccupy philosophers of science today. Nevertheless, *The New Organon* remains a work of extraordinary intellectual daring – a challenge to the entire edifice of contemporary philosophy and learning. It has left its mark on all subsequent philosophical discussions of scientific method and has shaped accounts of the development, from the seventeenth century onwards, of so-called English empiricism. Bacon's symbolic role as a philosophical founding father, based on his groundbreaking inductive method, is still aptly summed up in the celebratory ode with which Abraham Cowley prefaced Thomas Sprat's early *History of the Royal Society* (1667):

> From these and all long Errors of the way,
> In which our wandring Praedecessors went ...
> Bacon, like Moses, led us forth at last,
> The barren Wilderness he past,
> Did on the very Border stand
> Of the blest promis'd Land,
> And from the Mountain Top of his Exalted Wit,
> Saw it himself, and shewed us it.

Science and politics

Sprat gives us the heroic, intellectually uncompromising Francis Bacon. In the end, however, Sir Francis Bacon, Earl Verulam, arch-pragmatist in affairs of state, was prepared to compromise even the intellectual rigour of his *New Organon* for political expediency.

In theory, all existing philosophical systems must be abandoned to make way for the one true inductive method. Still, Bacon relented when it came to the kinds of gentlemanly debates on intellectual matters in which he and his court friends engaged:

We do not in any way discourage these traditional subjects from generating disputations, enlivening discourse and being widely applied to professional use and the benefit of civil life, and from being accepted by general agreement as a kind of currency. Furthermore, we freely admit that our new proposals will not be very useful for those purposes, since there is no way that they can be brought down to the common understanding, except through their results and effects. But our

published writings (and especially the books On the Advancement of Learning)
testify how sincerely we mean what we say of our affection and goodwill towards
the accepted sciences. (I.128)

Perhaps this is what William Harvey had in mind as philosophy practised
with the political tact of a Lord Chancellor. Here Bacon gives in to those
old friends of his, including Henry Wotton and Toby Matthew, on whose
support and informed comment on his philosophical writings he relied,
but who were deeply sceptical of his desire to start the whole enterprise of
science again from first principles.

In 1607, for instance, Bacon's long-time associate and amanuensis
Thomas Bodley wrote a detailed set of notes for Bacon on his *Cogitata
et visa* – early notes towards *The New Organon*. Bodley expressed broad
scepticism at the idea that knowledge grounded in worldwide 'experience',
or observed reality, could be a substitute (rather than a supplement) for that
accumulated down through the ages. Suppose, he argued, we were 'first to
condemn our present knowledge of doubts and incertitude, and disclaim
all our axioms, maxims, and general assertions that are left by tradition
from our elders unto us, which (as it is to be intended) have passed all
probations of the sharpest wits that ever were'. Suppose we were to return,
as Bacon suggested, to an alphabet of nature, to rebuild science from first
principles – a task which would inevitably take centuries. Then, Bodley
concluded, we would be likely to find we had gone in a complete circle and
arrived back with the very science passed down to us by the ancients.[27]

The reader of *The New Organon* whom Bacon was keenest not to offend,
and for whose sake he was willing to make as many intellectual compro-
mises as necessary, was of course the king himself. James prided himself on
being familiar with fashionable scientific debate. He responded in person
to Bacon's dedications (both printed and private ones), graciously
acknowledging receipt of the presentation copy of *The New Organon*. He
intended to participate fully in the further refinement of the 'new instrument'. He was resolved, he wrote,

first, to read it through with care and attention, though I should steal some hours
from my sleep; having otherwise as little spare time to read it as you had to write it.
And then to use the liberty of a true friend, in not sparing to ask you the question
in any point whereof I shall stand in doubt: as, on the other part, I will willingly

---

[27] *Trecentale Bodleianum: A Memorial Volume for the Three Hundreth Anniversary of the Public Funeral
of Sir Thomas Bodley March 29, 1613* (Oxford, Clarendon Press, 1913), 145–63.

give a due commendation to such places as in my opinion shall deserve it. In the meantime, I can with comfort assure you, that you could not have made choice of a subject more befitting your place, and your universal and methodick knowledge; and in the general, I have already observed, that you jump with me, in keeping the midway between the two extremes; as also in some particulars I have found that you agree fully with my opinion.[28]

In his response, Bacon was understandably at pains to indicate that he took the king's offer entirely seriously. This would be a collaborative undertaking between the sovereign and his Lord Chancellor. And even though the guide to the inductive method was supposed to be sensory experience alone, Bacon would stretch a point and allow a special place for the intellectual observations of the king himself:

I cannot express how much comfort I received by your last letter of your own royal hand ... Your Majesty shall not only do to myself a singular favour, but to the business a material help, if you will be graciously pleased to open yourself to me in those things, wherein you may be unsatisfied. For though this work, as by position and principle, doth disclaim to be tried by anything but by experience, and the resultats of experience in a true way; yet the sharpness and profoundness of your Majesty's judgment ought to be an exception to this general rule; and your questions, observations, and admonishments, may do infinite good.

Besides, if the king was to assist with framing the new logic itself, perhaps he might be persuaded to put some financial backing behind the 'natural and experimental histories' which were ultimately to underpin the whole work:

This comfortable beginning makes me hope further, that your Majesty will be aiding to me, in setting men on work for the collecting of a natural and experimental history; which is *basis totius negotii* [grounds for the whole enterprise]; a thing which I assure myself will be from time to time an excellent recreation unto you; I say, to that admirable spirit of yours, that delighteth in light; and I hope well that even in your times many noble inventions may be discovered for man's use. For who can tell, now this mine of Truth is once opened, how the veins go, and what lieth higher and what lieth lower?[29]

No doubt James was suitably flattered at the idea that he might be competent to suggest refinements and modifications to the 'Great Instauration'. In private, however, he admitted candidly that Bacon's latest

---

[28] James to Francis Bacon, 16 October 1620. Spedding, *Letters and Life*, 7, 122.
[29] Francis Bacon to James, 20 October 1620. Ibid. 130–1.

effort was quite beyond his comprehension: 'His last book is like the peace of God, that passeth all understanding.' And as with so many of Bacon's projects, the hoped-for funding, the investment which would make the whole grand scientific enterprise possible, was not in the end forthcoming.

James Spedding, Bacon's devoted nineteenth-century editor and defender of his reputation, was at great pains to separate Bacon the philosopher and thinker from Bacon the political wheeler-dealer and time-server. In his authoritative edition, contextual material has been as far as possible removed from the philosophical and scientific works and is printed at a distance, in the *Letters and Life* volumes.[30] Within *The New Organon* we can, nevertheless, trace the inevitable interconnectedness of the two, and in the process arrive at a fuller understanding of Bacon's ground-breaking philosophy.

Lisa Jardine

[30] See Jardine and Stewart, 'Judge Him According to His Works'.

Chronology

	earl of Essex; Anthony joins him in 1592 after his return from France following the death of Elizabeth's spymaster, Walsingham
1589	MP for Liverpool (sponsored by Walsingham). Commissioned to write *An Advertisement Touching the Controversies of the Church of England* in response to the Marprelate controversy; Burghley grants Bacon the reversion of the Clerkship to the Council of the Star Chamber (a post worth a substantial annual sum, but which he did not actually obtain for twenty years)
1592	Commissioned to write *Observations upon a Libel* in response to a Jesuit anti-government invective; composes a performance piece for four speakers, 'Of Tribute', possibly as an entertainment at Gray's Inn, probably on Essex's behalf
1593	Anthony Bacon moves into Francis's chambers at Gray's Inn. Essex joins the Privy Council. Francis speaks in Parliament, opposing the granting of a subsidy to Queen Elizabeth; Burghley informs him that Elizabeth is furious. Subsequently she fails to promote him
1594	Becomes Learned Counsel (personal legal advisor) to Elizabeth; composes *Gesta Grayorum* for the Gray's Inn revels.
1595	Vigorous but unsuccessful campaign by Essex to have Bacon made Solicitor General. Bacon writes *Accession Day Device* on behalf of Essex for the queen's birthday on 17 November, but begins to distance himself publicly from the earl thereafter
1596	Writes *Maxims of the Law*
1597	First edition of Bacon's *Essays*, which have been circulating in manuscript, published together with his *Meditationes sacrae* and *Colours of Good and Evil*. Dedicated to Anthony Bacon but presented to Essex with an effusive private dedication, 'to whose disposition and commandment I have entirely and inviolably vowed my poor self, and whatever appertaineth unto me'
1601	Following Essex's rebellion, Bacon escapes prosecution when others in Essex's service are implicated and executed with him. Bacon is appointed one of the prosecutors at his trial for treason and writes *A Declaration of the Practices and Treasons Attempted and Committed by Robert, Late Earl of Essex*. Anthony Bacon is not prosecuted because of his ill health; he dies shortly thereafter
1603	Knighted by James I upon his accession to the throne after the

death of Elizabeth I; gains favour with the king as brother of the deceased Anthony, since Anthony, unlike Francis, had continued to support Essex to the end. Bacon writes *A Brief Discourse Touching the Happy Union of the Kingdoms of England and Scotland*

1604 Appointed Learned Counsel to the king

1605 Publishes *Two Bookes of the Proficiencie and Advancement of Learning, Divine and Humane* – generally known as *The Advancement of Learning* – a blueprint for how to improve the state of learning in the kingdom designed to flatter the king, who prided himself on his intellectual interests

1606 Marries fourteen-year-old Alice Barnham for her fortune (Bacon was forty-five); eyewitnesses describe the wedding of Bacon to 'his young wench' as a sumptuous affair

1607 Makes supremely eloquent speech to Parliament in favour of the union of Scotland and England; appointed Solicitor General

1609 *De sapientia veterum* (*Wisdom of the Ancients*) published

1610 Death of Bacon's mother, who had descended into mental infirmity some years earlier

1612 Publishes second and expanded edition of his popular *Essays*

1614 Appointed Attorney General; devises *Masque of Flowers* to honour the marriage of Robert Carr, earl of Somerset, and Frances Howard, performed on Twelfth Night

1616 Privy Councillor; presides at the successful trial of Frances Carr and Somerset for the murder of Thomas Overbury. Frances was subsequently pardoned, Somerset executed. Bacon gains the favour of the duke of Buckingham

1617 Bacon closely involved in the meteoric rise of Buckingham. Appointed Lord Keeper of the Great Seal; takes up residence in his father's house (and his own birthplace), York House in London

1618 Appointed Lord Chancellor in January. Presides over the successful prosecution and subsequent execution of Sir Walter Ralegh. Raised to the peerage as Baron Verulam in July

1619 Bacon and the king both ill with the stone. Queen Anne dies (James does not attend her funeral)

1620 Publishes first edition of *Instauratio magna* (*Great Instauration*, including *The New Organon*) in Latin in October

1621	Created Viscount St Albans in January. Impeached in May in the House of Lords for accepting bribes; briefly imprisoned at the end of May but released to house arrest on 2 June; retired to his house at Gorhambury on 23 June. Thereafter he returned only occasionally to York House in London to 'take physic and provide for his health'
1622	Publishes *History of Henry VII* in March (previously presented in manuscript to James in October 1621 in a bid to return to favour); publishes *History of Winds* and *History of Life and Death* in November (presenting the latter to Buckingham in January 1623)
1623	Publishes *De augmentis scientiarum*, the Latin translation and elaboration of *The Advancement of Learning*
1624	*Apophthegms* and *A Translation of Certain Psalms into English Verse* published
1625	King James dies and is succeeded by his son Charles I. Bacon officially dismissed from the Privy Council in April. Bacon publishes his further enlarged *Essayes, or Counsels, Civill and Morall*
1626	9 April, Bacon dies at the Highgate home of the earl of Arundel, having been taken violently ill on the return journey to Gorhambury after a visit to York House in London to take physic; 20 April, his widow Alice marries her late husband's gentleman usher, John Underhill Bacon's chaplain, William Rawley, publishes commemorative volume of Latin poems in Bacon's memory; Rawley begins posthumous printing of Bacon's unpublished papers: *Sylva sylvarum* and *New Atlantis* printed together with preface by Rawley
1629–58	Rawley publishes *Certain Miscellany Works* (1629), *Operum moralium et civilium tomus* (1638), *Resuscitatio* (with 'Life' by Rawley) (1657) and *Opuscula varia posthuma* (1658)

Further reading

The standard edition of Bacon's works is J. Spedding, R. L. Ellis and D. D. Heath (eds.), *Works* (7 vols., London, Longman et al., 1857–9). This contains Latin and English original texts and English translations of the major works. Spedding was concerned to separate from the works any possible taint derived from Bacon's infamous reputation as a politician and betrayer of friendship. He therefore allocated writings with any political or personal content to a separate 'Letters and Life', J. Spedding (ed.), *Letters and Life* (7 vols., London, Longman, Green, Longman & Roberts, 1861–74), which he prefaced with a lengthy apologia designed to exonerate Bacon. Many pieces in the *Letters and Life* belong more properly with the *Works*. For a full understanding of Bacon's writings it is necessary to use the two together. For English translations of additional minor works (the *Temporis partus masculus*, *Cogitata et visa* and *Redargutio philosophiarum*) see B. Farrington, *The Philosophy of Francis Bacon: An Essay on Its Development from 1603 to 1609* (Liverpool, Liverpool University Press, 1964). A text with a translation of Bacon's legal maxims, *Aphorismi de jure gentium maiore sive de fontibus justiciae et juris*, can be found in M. S. Neustadt, *The Making of the Instauration: Science, Politics and Law in the Career of Francis Bacon* (Ph.D. diss., Johns Hopkins University, 1987).

A new twelve-volume complete works is in progress, L. Jardine and G. Rees (general eds.), *The Oxford Francis Bacon* (Oxford, Oxford University Press, 1996–). It will reintegrate 'personal' and 'public' works, give facing-page translations, and add drafts and fragments which have come to light since Spedding. The first published volume of this edition contains philosophical studies forming part of the *Instauratio magna* which were not available to Spedding: G. Rees (ed.), *Philosophical Studies c.1611–c.1619*,

Oxford Francis Bacon 6 (Oxford, Clarendon Press, 1996). M. Kiernan (ed.), *Sir Francis Bacon: The Essayes or Counsels, Civill and Morall* (Oxford, Clarendon Press, 1986), will be included in the Oxford Francis Bacon. A companion Oxford edition of Bacon's correspondence is in progress: general editor A. Stewart.

For important work on the Latin text of *The New Organon* see M. Fattori, *Lessico del 'Novum Organum' di Francesco Bacone* (2 vols. Rome, Edizioni dell'Ateneo, 1980). There is a good French translation of *The New Organon*: M. Malherbe and J.-M. Pousseur (trans. and eds.), *Francis Bacon: Novum Organum* (Paris, Presses Universitaires de France, 1986).

Among the standard secondary works there are a number which established Bacon studies in the form in which we currently know them: P. Rossi, *Francis Bacon: From Magic to Science*, trans. Sacha Rabinovitch (London, Routledge & Kegan Paul, 1968); L. Jardine, *Francis Bacon: Discovery and the Art of Discourse* (Cambridge, Cambridge University Press, 1974); A. Pérez-Ramos, *Francis Bacon's Idea of Science and the Maker's Knowledge Tradition* (Oxford, Clarendon Press, 1988). F. H. Anderson, *The Philosophy of Francis Bacon* (Chicago, Chicago University Press, 1948), and B. Farrington, *Francis Bacon: Philosopher of Industrial Science* (New York, Henry Schuman, 1949), are still worth consulting. Most recently, P. Zagorin, *Francis Bacon* (Princeton, NJ, Princeton University Press, 1998), offers a useful synthesis.

The literature on the inductive method is vast, and partisan. Some classic treatments are to be found as follows: C. D. Broad, *The Philosophy of Francis Bacon* (Cambridge, Cambridge University Press, 1926); G. H. von Wright, *The Logical Problem of Induction* (2nd rev. ed., Oxford, Blackwell, 1957); M. B. Hesse, 'Francis Bacon's Philosophy of Science' in B. Vickers (ed.), *Essential Articles for the Study of Francis Bacon* (Hamden, CT, Archon Press, 1968; repr. 1972), 115–39; K. Popper, *The Logic of Scientific Discovery* (London, Hutchinson, 1977); L. J. Cohen, *The Philosophy of Induction and Probability* (Oxford, Clarendon Press, 1989). P. Urbach, *Francis Bacon's Philosophy of Science: An Account and a Reappraisal* (La Salle, IL, Open Court, 1987), contains a useful survey treatment of Bacon's inductive method.

Biographical work on Bacon has tended to be governed by the passionate desire of biographers to clear Bacon's name as a politician (not to mention the view of some that he might have authored Shakespeare's plays). The most complete biography is now L. Jardine and A. Stewart,

Hostage to Fortune: The Troubled Life of Francis Bacon (London, Gollancz, 1998).

On the fortunes of Bacon's writings in the period immediately following his death see C. Webster, *The Great Instauration: Science, Medicine, and Reform 1626-1660* (London, Duckworth, 1975). See also M. Hunter, *Science and Society in Restoration England* (Cambridge, Cambridge University Press, 1981).

Useful works on the literary nature of Bacon's writing include J. C. Briggs, *Francis Bacon and the Rhetoric of Nature* (Cambridge, MA, Harvard University Press, 1989); J. Stephens, *Francis Bacon and the Style of Science* (Chicago, University of Chicago Press, 1975); C. Whitney, *Francis Bacon and Modernity* (New Haven, CT, Yale University Press, 1986). A student edition of Bacon's literary works with commentaries has been published in the 'Oxford Authors' series: B. Vickers (ed.), *Francis Bacon* (Oxford, Oxford University Press, 1996).

On Bacon's history writing see S. Clark, 'Bacon's *Henry VII*: A Case-Study in the Science of Man', *History and Theory* 13 (1974), 97–118. On Bacon's writings on the law see D. R. Coquillette, *Francis Bacon* (Stanford, CA, Stanford University Press, 1992). On Bacon's politics see J. Martin, *Francis Bacon, the State, and the Reform of Natural Philosophy* (Cambridge, Cambridge University Press, 1992); J. G. Crowther, *Francis Bacon: The First Statesman of Science* (London, Cresset Press, 1960). On Bacon's ethics see K. R. Wallace, *Francis Bacon and the Nature of Man* (Urbana, University of Illinois Press, 1967).

Useful collections of essays on aspects of Bacon's thought are to be found in B. Vickers (ed.), *Essential Articles for the Study of Francis Bacon* (Hamden, CT, Archon Press, 1968; repr. 1972); M. Fattori (ed.), *Francis Bacon: terminologia e fortuna nel XVII secolo* (Rome, Edizioni dell'Ateneo, 1984); M. Malherbe and J.-M. Pousseur (eds), *Francis Bacon: science et méthode* (Paris, J. Vrin, 1985); W. A. Sessions (ed.), *Francis Bacon's Legacy of Texts: The Art of Discovery Grows with Discovery* (New York, AMS Press, 1990); M. Peltonen (ed.), *The Cambridge Companion to Bacon* (Cambridge, Cambridge University Press, 1996).

The Great Renewal

These are
the thoughts of
Francis Verulam,
and this is the
method which he designed for himself:
he believed
that present and future generations
would be better off
if he made it known to them.

He became aware that the human intellect is the source of its own problems, and makes no sensible and appropriate use of the very real aids which are within man's power; the consequence is a deeply layered ignorance of nature, and as a result of this ignorance, innumerable deprivations. He therefore judged that he must make every effort to find a way by which the relation between the mind and nature could be wholly restored or at least considerably improved. But there was simply no hope that errors which have grown powerful with age and which are likely to remain powerful for ever would (if the mind were left to itself) correct themselves of their own accord one by one, either from the native force of the understanding or with the help and assistance of logic. The reason is that the first notions of things which the mind accepts, keeps and accumulates (and which are the source of everything else), are faulty and confused and abstracted from things without care; and in its secondary and other notions there is no less passion and inconsistency. The consequence is that the general human reason which we bring to bear on the inquiry into nature is not well founded and properly constructed; it is like a magnificent palace without a foundation. Men admire and celebrate the false powers of the mind, but miss and lose the real powers they could have (if the proper assistance were used and if the mind itself were more compliant towards nature and did not recklessly insult it). The only course remaining was to try the thing again from the start with better means, and make a general Renewal of the sciences and arts and of all human learning, beginning from correct foundations. This might seem, on approach, to be something illimitably vast and beyond mortal strength, and yet in the treatment, will be found to be sane and sensible, more so than what has been done in the past. For one can see an end to it. Whereas in what is currently done in the sciences, there is a kind of giddiness, a perpetual agitation and going in a circle. He is also very aware of the solitude in which this experiment moves, and how hard, how unbelievably difficult, it is to get people to believe in it. Nevertheless he felt that he should not fail himself or abandon his subject without attempting to travel the only road open to the human mind. For it is better to make a beginning of a thing which has a chance of an end, than to get caught up in things which have no end, in perpetual struggle and exertion. These ways of thought are analogous in some way to the two legendary paths of action: the one is steep and difficult at the beginning but ends in the open; the other, at first glance easy and downhill, leads to impassable, precipitous places.[1] He could not be sure when such things would occur to anyone again in the future; he was particularly moved by the argument that he had not so far found anyone who had applied his mind to similar thoughts; and therefore he decided to give to the public the first parts that he had been able to complete. His haste was not due to ambition but to anxiety; if in the human way of things, anything should happen to him, there would still be extant an outline and plan of the thing which he had conceived in his mind; there would still exist also some indication of his genuine concern for the good of the human race. Certainly he regarded every other ambition as lower than the work that he had in hand. For either the matter in question is nothing, or it is so important that it should rightly be content with itself and not seek any external reward.

[1] The reference is to the 'choice of Heracles' told in e.g. Xenophon, *Memorabilia*, II.21.

To our most serene
and powerful
Prince and Lord,

James

by the Grace of God
King of Great Britain, France and Ireland,
Defender of the Faith, etc.[2]

Most serene and powerful King,
Your Majesty may perhaps charge me with theft for stealing from your affairs the time I needed for this work. I have no answer. One cannot restore time; unless perhaps the time which I took from your affairs may redound to the memory of your name and the honour of your age, if this work has any value. It is certainly quite new; a totally new kind of thing; though drawn from a very old model, namely the world itself, and the nature of things and of the mind. Certainly I myself (I frankly confess) am accustomed to regard this work as a birth of time rather than of intelligence. The only wonder is that the beginning of the thing and such a powerful suspicion of opinions so long prevalent could have entered anyone's mind. The rest follows freely. But undoubtedly chance (as we say) and a certain fortuitous element plays a role in what men think no less than in what they do or say. By this chance of which I speak, I mean that if there is any good in these things which I bring, it will be imputed to the immense mercy and goodness of God and to the happiness of your times: as I have served you in my life with the sincerest devotion, so after my death I may perhaps ensure that your age will shine to posterity, by the lighting of this new torch in the dark days of philosophy. And this Regeneration and Renewal of the sciences is rightly due to the times of the wisest and most learned of all kings. I would add a petition, not unworthy of your Majesty, and most closely related to our present subject. It is that as you rival Solomon[3] in so many things, in gravity of judgement, in the peace of your kingdom, in the largeness of your heart, and finally in the remarkable variety of books which you have composed, you would emulate that same king in another way, by taking steps to ensure that a Natural and Experimental History be

---

[2] James I reigned 1603–25. Francis Bacon served him in various high ministerial positions which culminated in a term as Lord Chancellor 1618–21.

[3] King of the ancient Hebrews (*c.* 937–932 BC); the wisdom of Solomon is proverbial.

built up and completed: the true, strict history (without philological questions) which is the path to the foundation of philosophy, and which we shall describe in its place. So that at last, after so many ages of the world, philosophy and the sciences may no longer float in the air, but rest upon the solid foundations of every kind of experience properly considered. I have supplied the Instrument;[+] but the material must be sought in things themselves. May the Great and Good God long preserve your Majesty from harm.

> *Your Serene Majesty's*
> *most faithful and*
> *devoted Servant,*

FRANCIS VERULAM,
CHANCELLOR

[+] 'Instrument' translates *Organum (=* Greek *Organon)* as in Bacon's title *Novum Organum,* which is literally 'The New Instrument'.

Francis
Verulam

The Great Renewal

Preface

On the state of the sciences, that it is neither prosperous nor far advanced;
and that a quite different way must be opened up for the human
intellect than men have known in the past, and
new aids devised, so that
the mind may exercise
its right over nature.

Men seem to me to have no good sense of either their resources or their power; but to exaggerate the former and underrate the latter. Hence, either they put an insane value on the arts which they already have and look no further or, undervaluing themselves, they waste their power on trifles and fail to try it out on things which go to the heart of the matter. And so they are like fatal pillars of Hercules[5] to the sciences; for they are not stirred by the desire or the hope of going further. Belief in abundance is among the greatest causes of poverty; because of confidence in the present, real aids for the future are neglected. It is therefore not merely useful but quite essential that at the very outset of our work (without hesitation or pretence) we rid ourselves of this excess of veneration and regard, with a useful warning that men should not exaggerate or celebrate their abundance and its usefulness. For if you look closely at the wide range of books which are the boast of the arts and sciences, you will frequently find innumerable repetitions of the same thing, different in manner of treatment but anticipated in content, so that things which at first glance seem to be numerous are found on examination to be few. One must also speak plainly about usefulness, and say that the wisdom which we have drawn in particular from the Greeks seems to be a kind of childish stage of science, and to have the child's characteristic of being all too ready to talk, but too weak and immature to produce anything. For it is fertile in controversies, and feeble in results. The story of Scylla seems to fit the current state of letters exactly: she showed the face and visage of a virgin, but barking monsters clothed

---

[5] *Columnae* (pillars) seems to allude to the engraving on the title page of the edition of 1620, and to refer to the Pillars of Hercules, beyond which men had not dared to sail hitherto. Cf. Plato, *Timaeus* 24D ff on the pillars of Hercules and Atlantis.

and clung to her loins.[6] Similarly, the sciences to which we are accustomed have certain bland and specious generalities, but when we get to particulars (which are like the generative parts), so that they may bring forth fruit and works from themselves, disputes and scrappy controversies start up, and that is where it ends and that is all the fruit they have to show. Besides, if such sciences were not a completely dead thing, it seems very unlikely that we would have the situation we have had for many centuries, that the sciences are almost stopped in their tracks, and show no developments worthy of the human race. Very often indeed not only does an assertion remain a mere assertion but a question remains a mere question, not resolved by discussion, but fixed and augmented; and the whole tradition of the disciplines presents us with a series of masters and pupils, not a succession of discoverers and disciples who make notable improvements to the discoveries. In the mechanical arts we see the opposite situation. They grow and improve every day as if they breathed some vital breeze. In their first authors they usually appear crude, clumsy almost, and ungainly, but later they acquire new powers and a kind of elegance, to the point that men's desires and ambitions change and fail more swiftly than these arts reach their peak of perfection. By contrast, philosophy and the intellectual sciences are, like statues, admired and venerated but not improved. Moreover they are sometimes at their best in their earliest author and then decline. For after men have joined a sect and committed themselves (like obsequious courtiers) to one man's opinion, they add no distinction to the sciences themselves, but act like servants in courting and adorning their authors. Let no one maintain that the sciences have grown little by little and now have reached a certain condition, and now at last (like runners who have finished the race) have found their final homes in the works of a few authors, and now that nothing better can be discovered, it remains only to adorn and cultivate what has already been discovered. We could wish that it were so. But a more correct and truthful account of the matter is that these appropriations of the sciences[7] are simply a result of the confidence of a few men and the idleness and inertia of the rest. For after the sciences had been perhaps carefully cultivated and developed in some areas, by chance there arose a person, daring in character, who was accepted and followed because he had a summary kind of method; in appearance he gave the art a form, but in reality he corrupted the labours of the older

[6] For this portrait of Scylla see Ovid, *Metamorphoses*, XIII.732–3.
[7] 'this appropriating of the sciences' (Ellis)

7

investigators. Yet it is a delight to posterity, because of the handy usefulness of his work and their disgust and impatience with new inquiry. And if
anyone is attracted by ancient consensus and the judgement of time (so to
speak), he should realise that he is relying on a very deceptive and feeble
method. For we are mostly ignorant of what has become known and been
published in the sciences and arts in different centuries and other places,
and much more ignorant of what has been tried by individuals and discussed in private. So neither the births nor the abortions of time are extant
in the public record. Nor should we attach much value to consensus itself
and its longevity. There may be many kinds of political state, but there is
only one state of the sciences, and it is a popular state and always will be.
And among the people the kinds of learning which are most popular are
those which are either controversial and combative or attractive and empty,
that is, those which ensnare and those which seduce assent. This is surely
why the greatest geniuses in every age have suffered violence; while men of
uncommon intellect and understanding, simply to preserve their reputation, have submitted themselves to the judgement of time and the multi-
tude. For this reason, if profound thoughts have occasionally flared up, they
have soon been blown on by the winds of common opinion and put out.
The result is that Time like a river has brought down to us the light things
that float on the surface, and has sunk what is weighty and solid. Even those
authors who have assumed a kind of dictatorship in the sciences and make
pronouncements about things with so much confidence, take to complain-
ing when they recover their senses from time to time about the subtlety of
nature, the depths of truth, the obscurity of things, the complexity of
causes, and the weakness of human understanding; yet they are no more
modest in this, since they prefer to blame the common condition of man
and nature rather than admit their own incapacity. In fact their usual habit,
when some art fails to deliver something, is to declare the thing impossible
on the basis of the same art. An art cannot be condemned when it is itself
both the advocate and the judge; and so the issue is to save ignorance from
disgrace. This then, more or less, is the condition of the traditional and
received kinds of learning: barren of results, full of questions; slow and
feeble in improvement; claiming perfection in the whole, but very imperfect
in the parts; popular in choice and suspect to the authors themselves, and
therefore wrapped up and presented with a variety of devices. Even those
who have set out to learn for themselves and to commit themselves to the
sciences and extend their limits, have not dared to abandon the received

sciences completely or to seek the sources of things. They think they have achieved something important if they insert and add something of their own, prudently reflecting that in assenting they preserve their modesty and in adding they keep their freedom. But in being respectful of opinions and habits, these middle ways that people praise result in great losses for the sciences. For you can hardly admire an author and at the same time go beyond him. It is like water; it ascends no higher than its starting point. And so such men make some emendations but little progress; they improve existing learning but do not progress to anything new. There have also been men who with greater daring have thought that everything was new with them, and have relied on the strength of their genius to flatten and destroy everything that went before, and so made room for themselves and their opinions. They have not achieved much for all their noise; for what they tried to do was not to augment philosophy and the arts in fact and effect, but only to cause a change in belief and transfer the leadership of opinion to themselves; with very little profit, since among opposite errors, the causes of erring are almost the same. Those who have had sufficient spirit to want other men to join their inquiries, because they were not enslaved to their own or to other people's dogmas but favoured freedom, have doubt-less been honest in intention, but they have been ineffective in practice. For they seem to have followed only probable reasoning, and are carried round and round in a whirlpool of arguments, and take all the power out of their investigation by their undisciplined licence in raising questions. There has been no one who has spent an adequate amount of time on things them-selves and on experience. And some again who have committed themselves to the waves of experience, making themselves almost mechanics, still practise a kind of aimless investigation in experience itself, since even they do not work by fixed rules. In fact most of them have set themselves some petty tasks, thinking it a great achievement to make a single discovery; a design as inept as it is modest. It is impossible to make a thorough and successful inquiry into the nature of a thing in the thing itself; after a tedious variety of experiments he finds no end but only further lines of investigation. Then again, one should particularly notice that every effort expended on experience right from the beginning has sought to obtain certain specific results and to get them fast and directly; it has sought (I repeat) profitable, not illuminating, experiments; failing to imitate God's order, who on the first day created only light, and devoted a whole day to it; and produced on that day no material effects, moving on to these only on subsequent days. But those who have assigned the highest functions to logic and have thought to fashion the most powerful assistants to the sciences out of logic, have well and truly seen that the unaided human understanding really has to be distrusted. However, the medicine is much worse than the disease; and not without its own problems. For the logic now in use, though very properly applied to civil questions and the arts which consist of discussion and opinion, still falls a long way short of the subtlety of nature; and in grasping at what it cannot hold, has succeeded in establishing and fixing errors rather than in opening up the way to truth.

And so, to summarise what I have said, neither a man's own efforts nor his trust in another's seems so far to have worked for men in the sciences; especially as there is little help to be got from the demonstrations and experiments so far known. The fabric of the universe, its structure, to the mind observing it, is like a labyrinth, where on all sides the path is so often uncertain, the resemblance of a thing or a sign is deceptive, and the twists and turns of natures are so oblique and intricate. One must travel always through the forests of experience and particular things, in the uncertain light of the senses, which is sometimes shining and sometimes hidden. Moreover those who offer to guide one on the way are also lost in the labyrinth and simply add to the number who have gone astray. In such difficult circumstances, one cannot count on the unaided power of men's judgement; one cannot count on succeeding by chance. Even supreme intelligence or unlimited throws of the dice could not overcome the difficulties. We need a thread to guide our steps; and the whole road, right from the first perceptions of sense, has to be made with a sure method. This should not be taken to imply that nothing at all has been achieved in so many centuries, with so much effort. Nor do we complain of the discoveries that have been made. Certainly in the things that were within the range of their intelligence and abstract thinking, the ancients acquitted themselves admirably. But just as in previous centuries when men set their course in sailing simply by observations of the stars, they were certainly able to follow the shores of the old continent and cross some relatively small inland seas, but before the ocean could be crossed and the territories of the new world revealed, it was necessary to have a knowledge of the nautical compass as a more reliable and certain guide. By the same reasoning exactly, the discoveries that have so far been made in the arts and sciences are of the kind that could be found out by use, thought, observation and argument, in that they are closely connected with the senses and common notions; but before one can sail to the more remote and secret places of nature, it is absolutely essential to introduce a better and more perfect use and application of the mind and understanding.

For ourselves, swayed by the eternal love of truth, we have committed ourselves to uncertain, rough and solitary ways, and relying and resting on God's help, we have fortified our mind against violent attacks from the armed forces of opinion, and against our own internal hesitations and scruples, the dark mists and clouds and fantasies of things flying all around us; so that at the end we may be able to provide more reliable and secure directions[8] for present and future generations. If we have had any success in this, the method that opened the way for us was certainly a true and proper humiliation of the human spirit. For all those before us who have devoted themselves to the discovery of arts have simply cast a brief glance at things and examples and experience, and then called on their own spirits to give them oracles, as if discovery were no more than conjuring up a new idea. But we stay faithfully and constantly with things, and abstract our minds no further from them than is necessary for the images and rays of things to come into focus (as in the case of sight), and therefore little is left to the power and excellence of the intelligence. And as we use humility in discovery, we have followed it also in teaching. And we do not attempt to claim or impose a spurious dignity on our discoveries either by triumphs in refutation or by appeals to antiquity or by any usurpation of authority or even by taking refuge in obscurity; it would not be difficult to do this kind of thing if one were trying to glorify his own name rather than enlighten the minds of others. We have not planned (I say) or laid any attack or ambush for men's judgements; we bring them into the presence of things themselves and their connections, so that they may see what they have, what they may question, and what they may add and contribute to the common stock. If we have too readily believed anything, if we have fallen asleep or not paid enough attention, or given up on the way and stopped the inquiry too soon, we still present things plainly and clearly. Hence our mistakes may be noted and removed before they infect the body of science too deeply; and anyone else may easily and readily take over our labours. In this way we believe that we have made for ever a true and lawful

---

[8] *indicia*: cf. the full Latin title of the *New Organon*: *Novum Organum, sive Indicia vera de interpretatione naturae*, i.e. 'The New Instrument, or True Directions for the Interpretation of Nature'.

marriage between the empirical and the rational faculties (whose sad and unhappy divorce and separation have caused all the trouble in the human family).

And therefore, since these things are not under our control, at the outset of our work we offer the most humble and fervent prayers to God the Father, God the Word and God the Spirit, that mindful of the afflictions of mankind and of the pilgrimage of life in which we pass few days and evil, they may deign to endow the human family through our hands with new mercies. We also humbly pray that the human may not overshadow the divine, and that from the revelation of the ways of sense and the brighter burning of the natural light, the darkness of unbelief in the face of the mysteries of God may not arise in our hearts. Rather we pray that from a clear understanding, purged of fantasy and vanity, yet subject still to the oracles of God and wholly committed to them, we may give to faith all that belongs to faith. And finally we pray that when we have extracted from knowledge the poison infused by the serpent which swells and inflates the human mind, we may not be wise with too high or too great a wisdom, but may cultivate the truth in all charity.

Our prayers done, we turn to men and offer some salutary advice and make some reasonable requests. First we advise (as we have prayed) that men may restrain their sense within their duty, so far as the things of God are concerned. For sense (like the sun) opens up the face of the terrestrial globe and closes and obscures the globe of heaven. And then we warn men not to err in the opposite direction as they avoid this evil; which will certainly happen if they believe that any part of the inquiry into nature is forbidden by an interdict. The pure and immaculate natural knowledge by which Adam assigned appropriate names to things did not give opportunity or occasion for the Fall. The method and mode of temptation in fact was the ambitious and demanding desire for moral knowledge, by which to discriminate good from evil, to the end that Man might turn away from God and give laws to himself. About the sciences which observe nature the sacred philosopher declares that 'the Glory of God is to conceal a thing, but the glory of a king is to find out a thing',[9] just as if the divine nature delighted in the innocent and amusing children's game in which they hide themselves purposely in order to be found; and has coopted the human mind to join this game in his kindness and goodness towards men. Finally,

---

[9] Proverbs 25:2; the phrase is quoted again at *New Organon*, I.129, in a slightly different form.

we want all and everyone to be advised to reflect on the true ends of knowledge:[10] not to seek it for amusement or for dispute, or to look down on others, or for profit or for fame or for power or any such inferior ends, but for the uses and benefits of life, and to improve and conduct it in charity. For the angels fell because of an appetite for power; and men fell because of an appetite for knowledge; but charity knows no bounds; and has never brought angel or man into danger.

The requests we make are as follows. Nothing for ourselves personally, but about what we are doing, we ask that men think of it not as an opinion but as a work, and hold it for certain that we are laying the foundations not of a sect or of a dogma, but of human progress and empowerment. And then that they would give their own real interests a chance, and put off the zeal and prejudice of beliefs and think of the common good; then, freed from obstacles and mistaken notions of the way, and equipped with our helps and assistance, we would ask them to undertake their share of the labours that remain. And we ask them to be of good hope; and not imagine or conceive of our *Renewal* as something infinite and superhuman, when in fact it is the end of unending error, and the right goal, and accepts the limitations of mortality and humanity, since it does not expect that the thing can be completely finished in the course of one lifetime, but provides for successors; and finally that it seeks knowledge not (arrogantly) in the tiny cells of human intelligence but humbly in the wider world. For the most part empty things are very big, solid things are very dense and take up little space. Finally, it seems, we must also request (just in case anyone means to be unfair to us, which would imperil the project itself) that men determine how far, on the basis of what we are compelled to say (if we are to be consistent), they may believe they have the right to have an opinion or to express a view about our teachings; for we reject (in an inquiry into nature) all that hasty human reasoning, based on preconceptions,[11] which abstracts from things carelessly and more quickly than it should, as a vague, unstable procedure, badly devised. And I cannot be arraigned to stand trial under a procedure which is itself on trial.

[10] *scientia*
[11] *anticipantem*: see I.26 on 'anticipations of nature'.

The plan of the work

It consists of six Parts:

First, *The Divisions of the Sciences.*
Second, *The New Organon,* or *Directions for the Interpretation of Nature.*
Third, *Phenomena of the Universe,* or *A Natural and Experimental History towards the foundation of Philosophy.*
Fourth, *The Ladder of the Intellect.*
Fifth, *Forerunners,* or *Anticipations of Second Philosophy.*
Sixth, *Second Philosophy,* or *Practical Science.*

The outlines of each Part

It is a part of our plan to set everything out as openly and clearly as possible. For a naked mind is the companion of innocence and simplicity, as once upon a time the naked body was. And therefore we must first lay out the order and plan of our work. It consists of six parts.

The first part gives a Summary or general description of the science or learning which the human race currently possesses. It seemed good to us to spend some time on what is presently accepted, thinking that this would help the perfection of the old and the approach to the new. We are almost equally eager to develop the old and to acquire the new. This also gives us credibility, according to the saying that 'an ignorant man will not believe words of knowledge until you have told him what he has in his heart'. Hence we shall not neglect to sail along the shores of the accepted sciences and arts, importing some useful items into them, in our passage.

However the divisions of the sciences which we employ include not only things which have been noticed and discovered but also things that until now have been missed but should be there. For in the intellectual as in the physical world, there are deserts as well as cultivated places. And so it is not surprising if we sometimes depart from the customary divisions. An addition not only changes the whole, but necessarily also alters the parts and sections; and the accepted divisions merely reflect the currently accepted outline of the sciences.

In matters which we shall note as missing, we shall be sure to do more than simply suggest a bare title and an outline account of what is needed. For if we report among things missing anything (of some value) whose method seems so obscure that we are justified in suspecting that men will not easily understand what we mean, or what is the task which we imagine and conceive in our mind, we will always take the trouble either to add instructions for carrying out the task or a report of our own performance of a part of it, as an example of the whole; so that we may give some help in each case either by advice or in practice. We feel that our own reputation, as well as the interest of others, requires that no one should suppose that some superficial notions on these matters have simply entered in our heads, and that the things we desiderate and try to grasp are mere wishes. They are such that they are clearly within men's power (unless men fail themselves), and I do have a firm and explicit conception of them. I have undertaken not merely to survey these regions in my mind, like an augur taking the auspices, but to enter them like a general, with a strong will to claim possession. *And this is the first part of the work.*

After coasting by the ancient arts, we will next equip the human understanding to set out on the ocean. We plan therefore, for our second part, an account of a better and more perfect use of reason in the investigation of things and of the true aids of the intellect, so that (despite our humanity and subjection to death) the understanding may be raised and enlarged in its ability to overcome the difficult and dark things of nature. And the art which we apply (which we have chosen to call *Interpretation of Nature*) is an art of logic, though with a great difference, indeed a vast difference. It is true that ordinary logic also claims to devise and prepare assistants and supports for the intellect; in this they are the same. But it differs altogether from ordinary logic in three particular ways: viz., in its end, in its order of demonstration, and in the starting points of its inquiry.

For the end we propose for our science is the discovery of arts, not of arguments, of principles and not of inferences from principles, of signs and indications of works and not probable reasonings. Different results follow from our different design. They defeat and conquer their adversary by disputation; we conquer nature[12] by work.

The nature and order of our demonstrations agree with such an end. For in ordinary logic almost all effort is concentrated on the syllogism. The logicians seem scarcely to have thought about induction. They pass it by with barely a mention, and hurry on to their formulae for disputation. But we reject proof by syllogism, because it operates in confusion and lets nature slip out of our hands. For although no one could doubt that things which agree in a middle term, agree also with each other (which has a kind of mathematical certainty), nevertheless there is a kind of underlying fraud here, in that a syllogism consists of propositions, and propositions consist of words, and words are counters and signs of notions. And therefore if the very notions of the mind (which are like the soul of words, and the basis of every such structure and fabric) are badly or carelessly abstracted from things, and are vague and not defined with sufficiently clear outlines, and thus deficient in many ways, everything falls to pieces. And therefore we reject the syllogism; and not only so far as principles are concerned (they do not use it for that either) but also for intermediate propositions, which the syllogism admittedly deduces and generates in a certain fashion, but without effects, quite divorced from practice and completely irrelevant to the active part of the sciences. For even if we leave to the syllogism and similar celebrated but notorious kinds of demonstration jurisdiction over the popular arts which are based on opinion (for we have no ambitions in this area), still for the nature of things we use induction throughout, and as much for the minor propositions as for the major ones. For we regard *induction* as the form of demonstration which respects the senses, stays close to nature, fosters results and is almost involved in them itself.

And so the order of demonstration also is completely reversed. For the way the thing has normally been done until now is to leap immediately from sense and particulars to the most general propositions, as to fixed poles around which disputations may revolve; then to derive everything else from them by means of intermediate propositions; which is certainly

[12] Reading *natura*, for the *naturá* of the edition of 1620. This is the reading which Kitchin's translation presupposes. (See *The Novum Organum; or, A True Guide to the Interpretation of Nature, by Francis Bacon, Lord Verulam. A New Translation by the Rev. G.W. Kitchin* (Oxford University Press, 1855)).

a short route, but dangerously steep, inaccessible to nature and inherently prone to disputations. By contrast, by our method, axioms are gradually elicited step by step, so that we reach the most general axioms only at the very end; and the most general axioms come out not as notional, but as well defined, and such as nature acknowledges as truly known to her, and which live in the heart of things.

By far the biggest question we raise is as to the actual form of induction, and of the judgement made on the basis of induction. For the form of induction which the logicians speak of, which proceeds by simple enumeration, is a childish thing, which jumps to conclusions, is exposed to the danger of instant contradiction, observes only familiar things and reaches no result.

What the sciences need is a form of induction which takes experience apart and analyses it, and forms necessary conclusions on the basis of appropriate exclusions and rejections. And if the logicians' usual form of judgement has been so difficult and required so much intellectual exertion, how much more effort should we expend on this other judgement, which is drawn not only from the depths of the mind but from the bowels of nature.

And this is not all. For we place the foundations of the sciences deeper and lay them lower, and set our starting points further back than men have ever done before, subjecting them to examination, while ordinary logic accepts them on the basis of others' belief. For logicians borrow (if I may put it this way) the principles of the sciences from the particular sciences themselves; then they pay respect to the first notions of the mind; finally they are happy with the immediate perceptions of healthy senses. But our position is that true logic should enter the provinces of the individual sciences with greater authority than is in their own principles, and compel those supposed principles themselves to give an account as to what extent they are firmly established. As for the first notions of the intellect: not one of the things which the intellect has accumulated by itself escapes our suspicion, and we do not confirm them without submitting them to a new trial and a verdict given in accordance with it. Furthermore, we have many ways of scrutinising the information of the senses themselves. For the senses often deceive, but they also give evidence of their own errors; however the errors are to hand, the evidence is far to seek.

The senses are defective in two ways: they may fail us altogether or they may deceive. First, there are many things which escape the senses even when they are healthy and quite unimpeded; either because of the rarity of the whole body or by the extremely small size of its parts, or by distance, or by its slowness or speed, or because the object is too familiar, or for other reasons. And even when the senses do grasp an object, their apprehensions of it are not always reliable. For the evidence and information given by the senses is always based on the analogy of man not of the universe; it is a very great error to assert that the senses are the measure of things.

So to meet these defects, we have sought and gathered from every side, with great and faithful devotion, assistants to the senses, so as to provide substitutes in the case of total failure and correction in the case of distortion. We do this not so much with instruments as with experiments. For the subtlety of experiments is far greater than that of the senses themselves even when assisted by carefully designed instruments; we speak of experiments which have been devised and applied specifically for the question under investigation with skill and good technique. And therefore we do not rely very much on the immediate and proper perception of the senses, but we bring the matter to the point that the senses judge only of the experiment, the experiment judges of the thing. Hence we believe that we have made the senses (from which, if we prefer not to be insane we must derive everything in natural things) sacred high priests of nature and skilled interpreters of its oracles; while others merely seem to respect and honour the senses, we do so in actual fact. Such are the preparations which we make for the light of nature and its kindling and application; and they would be sufficient in themselves if men's understandings were unbiased, a blank slate. But as men's minds have been occupied in so many strange ways that they have no even, polished surface available to receive the true rays of things, it is essential for us to realise that we need to find a remedy for this too.

The *Idols*[13] by which the mind is occupied are either artificial or innate. The artificial *idols* have entered men's minds either from the doctrines and sects of philosophers or from perverse rules of proof. The innate idols are inherent in the nature of the intellect itself, which is found to be much more prone to error than the senses. For however much men may flatter themselves and run into admiration and almost veneration of the human mind,

---

[13] 'Idols' is the usual translation of Bacon's famous *idola*. We too have used it on most occasions, but the meaning of some passages seemed to be better conveyed by the translation 'illusion'. See also 1.39n.

it is quite certain that, just as an uneven mirror alters the rays of things from their proper shape and figure, so also the mind, when it is affected by things through the senses, does not faithfully preserve them, but inserts and mingles its own nature with the nature of things as it forms and devises its own notions.

The first two kinds of *idols* can be eliminated, with some difficulty, but the last in no way. The only strategy remaining is, on the one hand, to indict them, and to expose and condemn the mind's insidious force, in case after the destruction of the old, new shoots of error should grow and multiply from the poor structure of the mind itself, and the result would be that errors would not be quashed but simply altered; and on the other hand, to fix and establish for ever the truth that the intellect can make no judgement except by induction in its legitimate form. Hence the teaching which cleanses the mind to make it receptive to truth consists of three refutations: a refutation of philosophies; a refutation of proofs; and a refutation of natural human reason. When we have dealt with these, and clarified the part played by the nature of things and the part played by the nature of the mind, we believe that, with the help of God's goodness, we will have furnished and adorned the bedchamber for the marriage of the mind and the universe. In the wedding hymn we should pray that men may see born from this union the assistants that they need and a lineage of discoveries which may in some part conquer and subdue the misery and poverty of man. *And this is the second part of the work.*

But we plan not only to show the way and build the roads, but also to enter upon them. And therefore the third part of our work deals with the *Phenomena of the Universe*, that is, every kind of experience, and the sort of natural history which can establish the foundations of philosophy. A superior method of proof or form of interpreting nature may defend and protect the mind from error and mistake, but it cannot supply or provide material for knowledge. But those who are determined not to guess and take omens but to discover and know, and not to make up fairytales and stories about worlds, but to inspect and analyse the nature of this real world, must seek everything from things themselves. No substitute or alternative in the way of intelligence, thought or argument can take the place of hard work and investigation and the visitation of the world, not even if all the genius of all the world worked together. This then we must unfailingly do or abandon the business for ever. But to this very day men have acted so foolishly that it is no wonder that nature does not give them access to her.

For in the first place, the information of the senses themselves is defective and deceiving; observation is lazy, uneven and casual; teaching is empty and based on hearsay; practice is slavishly bent on results; experimental initiative is blind, unintelligent, hasty and erratic; and natural history is shallow and superficial. Between them they have accumulated very poor material for the intellect to construct philosophy and the sciences.

And the tendency to introduce subtle and intricate disputation prematurely comes too late to remedy a situation which is utterly desperate, and does nothing to move on the enterprise or remove error. Thus there is no hope of major development or progress except in a renewal of the sciences.

Its beginnings must come from a natural history, and a natural history of a new kind with a new organisation. It would be pointless to polish the mirror if there were no images; and clearly we must get suitable material for the intellect, as well as making reliable instruments. And our history (like our logic) differs from that now in use in many ways: in its purpose or task, in its actual extent and composition, in its subtlety, and also in the selection and arrangement of it in relation to the next stage.

First we propose a natural history which does not so much amuse by the variety of its contents or give immediate profit from its experiments, as shed light on the discovery of causes and provide a first breast to feed philosophy. For although our ultimate aim is works and the active part of science, still we wait for harvest time and do not try to reap moss and the crop while it is still green. We know very well that axioms properly discovered bring whole companies of works with them, revealing them not singly but in quantity. But we utterly condemn and reject the childish desire to take some pledges prematurely, in the form of new works, like an apple of Atalanta which slows the race.[14] Such is the task of our Natural History.

And as for its composition, we are making a history not only of nature free and unconstrained (when nature goes its own way and does its own work), such as a history of the bodies of heaven and the sky, of land and sea, of minerals, plants and animals; but much more of nature confined and harassed, when it is forced from its own condition by art and human

---

[14] This was one of the golden balls (or apples) which Milanion threw in front of Atalanta while he was racing her, so that he could win the race and her hand in marriage.

agency, and pressured and moulded. And therefore we give a full description of all the experiments of the mechanical arts, all the experiments of the applied part of the liberal arts, and all the experiments of several practical arts which have not yet formed a specific art of their own (so far as we have had an opportunity to investigate and they are relevant to our purpose). Moreover (to be plain) we put much more effort and many more resources into this part than into the other, and pay no attention to men's disgust or what they find attractive, since nature reveals herself more through the harassment of art than in her own proper freedom.

And we do not give a history of bodies only; we felt that we should also take the trouble to make a separate history of the powers themselves (we mean those which could be considered as central powers in nature, and which plainly constitute the originals of nature, since they are the material for the first passions and desires, viz., *Dense, Rare, Hot, Cold, Solid, Liquid, Heavy, Light* and many others).

As for subtlety, we are certainly looking for a kind of experience which is far more subtle and simple than those which simply happen. For we bring and draw many things out of obscurity which no one would ever have thought to investigate if he were not following the sure and steady path to the discovery of causes. For in themselves they are of no great use, so that it is quite clear that they have not been sought for themselves. Rather they are to things and works exactly like the letters of the alphabet to speech and words: though useless in themselves, they are still the elements of all discourse.

And in the choice of narratives and experiences we think that we have served men better than those who have dealt with natural history in the past. For we use the evidence of our own eyes, or at least of our own perception, in everything, and apply the strictest criteria in accepting things; so that we exaggerate nothing in our reports for the sake of sensation, and our narrations are free and untouched by fable and foolishness. We also specifically proscribe and condemn many widely accepted falsehoods (which have prevailed for many centuries by a kind of neglect and are deeply ingrained), so that they may not trouble the sciences any more. For as someone wisely remarked that the stories and superstitions and trifles which nurses instil into children also seriously deprave their minds,[15] by the same reasoning we feel we must be careful, and even anxious, that

---

[15] Perhaps a reference to Plato, *Republic* 377A, 387B, *Laws* 793Dff, and elsewhere.

philosophy should not at the start get into the habit of any kind of foolishness as we foster and nurture its infancy in the form of natural history. In every experiment which is new and even the least bit subtle, even if (as it seems to us) it is sure and proven, we give a frank account of the method of the experiment we used; so that after we have revealed every move we made, men may see any hidden error attached to it, and may be prompted to find more reliable, more meticulous proofs (if any exist); and finally we sprinkle warnings, reservations and cautions in all directions, with the religious scruple of an exorcist casting out and banishing every kind of fantasy.

Finally, since we have seen how much experience and history distort the sight of the human mind, and how difficult it is (especially for tender or prejudiced minds) at first to get used to nature, we often add our own observations, which are like the first turn or move of history towards philosophy (perhaps one might say, the first glance). They are intended to be like a pledge to men that they will not be forever floundering in the waves of history, and that when we come to the work of the understanding, everything will be more ready for action. By such a Natural History (as we have outlined) we believe that men may make a safe, convenient approach to nature and supply good, prepared material to the understanding.

After[16] we have surrounded the intellect with the most trustworthy aides and bodyguards, and have used the most stringent selection to build a fine army of divine works, it may seem that nothing remains to be done but to approach philosophy itself. But in such a difficult and doubtful task there are certain points which it seems necessary to introduce first, partly for instruction and partly for their immediate usefulness.

The first point is to give examples of investigation and discovery by our way and method, as exhibited in certain subjects. We particularly choose subjects which are both the most notable of things under investigation and the most different from each other; so that in every *genus* we may have an example. We are not speaking of examples added to individual precepts and rules for illustration (these we have given in abundance in our second part); we simply mean types and variations, which may bring before our eyes the whole procedure of the mind and the seamless fabric and order of its discovery of things, in certain subjects, which will be diverse and striking.

---

[16] This is the beginning of the fourth part.

The analogy that suggests itself is that in mathematics demonstration is easy and clear when the machine is used, whereas without this convenience everything seems complicated and more subtle than it really is. And so we devote the *fourth part* of our work to such examples, and thus it is truly and simply a particular and detailed application of the second part.

The *fifth part* is useful only for a time until the rest is completed; and is given as a kind of interest until we can get the capital. We are not driving blindly towards our goal and ignoring the useful things that come up on the way. For this reason the fifth part of our work consists of things which we have either discovered, demonstrated or added, not on the basis of our methods and instructions for interpretation, but from the same intellectual habits as other people generally employ in investigation and discovery. For while we expect, from our constant converse with nature, greater things from our reflections than our intellectual capacity might suggest, these temporary results may in the meantime serve as shelters built along the road for the mind to rest in for a while as it presses on towards more certain things. However, we insist in the meantime that we do not wish to be held to these results themselves, because they have not been discovered or demonstrated by the true form of interpretation. One should not be frightened of such a suspension of judgement in a doctrine which does not assert simply that nothing can be known, but that nothing can be known except in a certain order and by a certain method; and meanwhile it has set up some degrees of certitude for use and comfort until the mind reaches its goal of explanation of causes. Nor were the schools of philosophers who simply maintained *lack of conviction*[17] inferior to those who claimed a freedom to make pronouncements. Yet the former did not provide assistance to the sense and understanding, as we have done, but totally undermined belief and authority; which is a very different thing and almost the opposite.

[6] Finally the *sixth* part of our work (which the rest supports and serves) at last reveals and expounds the philosophy which is derived and formed from the kind of correct, pure, strict inquiry which we have already framed and explained. It is beyond our ability and beyond our expectation to achieve this final part and bring it to completion. We have made a start on

---

[17] Bacon uses the Greek word *acatalepsia*, the mark of the ancient Sceptics. See I.37.

the task, a start which we hope is not despicable; the end will come from the fortune of mankind, such an end perhaps as in the present condition of things and the present state of thought men cannot easily grasp or guess. It is not merely success in speculation which is in question, but the human situation, human fortune and the whole potential of works. For man is nature's agent and interpreter; he does and understands only as much as he has observed of the order of Nature in work or by inference; he does not know and cannot do more.[18] No strength exists that can interrupt or break the chain of causes; and nature is conquered only by obedience. Therefore those two goals of man, *knowledge* and *power*, a pair of twins, are really come to the same thing, and works are chiefly frustrated by ignorance of causes.

The whole secret is never to let the mind's eyes stray from things themselves, and to take in images exactly as they are. May God never allow us to publish a dream of our imagination as a model of the world, but rather graciously grant us the power to describe the true appearance and revelation of the prints and traces of the Creator in his creatures.

And therefore, Father, you who have given visible light as the first fruits of creation and, at the summit of your works, have breathed intellectual light into the face of man, protect and govern this work, which began in your goodness and returns to your glory. After you had turned to view the works which your hands had made, you saw that all things were very good, and you rested. But man, turning to the works which his hands have made, saw that all things were vanity and vexation of spirit,[19] and has had no rest. Wherefore if we labour in your works, you will make us to share in your vision and in your sabbath. We humbly beseech that this mind may remain in us; and that you may be pleased to bless the human family with new mercies, through our hands and the hands of those others to whom you will give the same mind.

[18] Cf. i.i.
[19] Ecclesiastes, 1:14.

THE FIRST PART
OF THE
RENEWAL
WHICH CONTAINS
THE DIVISIONS OF THE SCIENCES
IS LACKING.

They may however to some extent be recovered from
The Second Book of The Proficience and
Advancement of Learning, Divine and Human

────────────

THERE FOLLOWS THE SECOND PART
OF THE
RENEWAL,
WHICH EXPLAINS THE ACTUAL ART
OF INTERPRETING NATURE AND OF THE
TRUE OPERATION
OF THE INTELLECT:
NOT IN THE FORM OF A REGULAR TREATISE,
BUT DIGESTED, IN SUMMARY FORM,
INTO APHORISMS.

THE SECOND PART OF THE WORK.

IT IS CALLED
THE NEW ORGANON
OR
TRUE DIRECTIONS
FOR THE INTERPRETATION
OF NATURE

Preface

Those who have presumed to make pronouncements about nature as if it were a closed subject, whether they were speaking from simple confidence or from motives of ambition and academical habits, have done very great damage to philosophy and the sciences. They have been successful in getting themselves believed and effective in terminating and extinguishing investigation. They have not done so much good by their own abilities as they have done harm by spoiling and wasting the abilities of others. Those who have gone the opposite way and claimed that nothing at all can be known, whether they have reached this opinion from dislike of the ancient sophists or through a habit of vacillation or from a kind of surfeit of learning, have certainly brought good arguments to support their position. Yet they have not drawn their view from true starting points, but have been carried away by a kind of enthusiasm and artificial passion, and have gone beyond all measure. The earlier Greeks however (whose writings have perished) took a more judicious stance between the ostentation of dogmatic pronouncements and the despair of *lack of conviction (acatalepsia)*;[20] and though they frequently complained and indignantly deplored the difficulty of investigation and the obscurity of things, like horses champing at the bit they kept on pursuing their design and engaging with nature; thinking it appropriate (it seems) not to argue the point (whether anything can be known), but to try it by experience. And yet they too, relying only on the impulse of the intellect, failed to apply rules, and staked everything on the mind's endless and aimless activity.

[20] See I.37 and note.

Our method, though difficult to practise, is easy to formulate. It is to establish degrees of certainty, to preserve sensation by putting a kind of restraint on it, but to reject in general the work of the mind that follows sensation; and rather to open and construct a new and certain road for the mind from the actual perceptions of the senses. This was certainly seen also by those who have given such an important role to logic. Clearly they sought assistance for the understanding and distrusted natural and spontaneous movements of the mind. But this remedy was applied too late, when the situation was quite hopeless, after daily habits of life had let the mind be hooked by hearsay and debased doctrine, and occupied by thoroughly empty *illusions*.[21] And so the art of logic took its precautions too late, and altogether failed to restore the situation; and has had the effect of fixing errors rather than of revealing truth. There remains one hope of salvation, one way to good health: that the entire work of the mind be started over again; and from the very start the mind should not be left to itself, but be constantly controlled; and the business done (if I may put it this way) by machines. If men had tackled mechanical tasks with their bare hands and without the help and power of tools, as they have not hesitated to handle intellectual tasks with little but the bare force of their intellects, there would surely be very few things indeed which they could move and overcome, no matter how strenuous and united their efforts. And if we might pause for a moment and look at an example, as if we were looking into a mirror, we might (if you please) ask the following: if an exceptionally heavy obelisk had to be moved to decorate a triumph or some such magnificent show, and men tackled it with their bare hands, would not a sensible spectator regard it as an act of utter lunacy? And all the more so if they increased the number of workers thinking that that would do it? Would he not say they were still more seriously demented if they proceeded to make a selection, and set aside the weaker men and took only the young and the strong, and expected to achieve their ambition that way? And if not satisfied even with this, they decided to have recourse to the art of athletics, and gave orders that everyone should turn up with hands, arms and muscles properly oiled and massaged according to the rules of the art, would he not protest that what they were doing was simply a systematic and methodical act of insanity? And yet in intellectual tasks men are motivated by a similarly insane impulse and an equally ineffective

---

[21] *idola*: for the translation of this term see 'Plan of the Work', n.13, and 1.39n.

enterprise when they expect much from either a cooperation of many minds or simple brilliance and high intelligence, or even when they improve the force of their minds with logic (which may be thought of as a kind of athletic art); and all the time, however much effort and energy they put into it (if one looks at it from a proper perspective), they are using nothing but the naked intellect. Yet it is utterly obvious that in any major work that the human hand undertakes, the strength of individuals cannot be increased nor the forces of all united without the aid of tools and machines.

From the premises given, we conclude that there are two things which we should like to bring to men's attention, so that they do not escape them or pass unnoticed. The first is this: by a happy chance (as we suppose) that tends to deflect and extinguish conceit and the spirit of contradiction, it is the case that we may carry out our design without touching or diminishing the honour and reverence due to the ancients, and still gather the fruit of our modesty. For if we maintained that we achieve better results than the ancients while following the same road as they, we should not by any skill with words be able to avoid setting up a comparison or contest in intellectual capacity or excellence. This by itself might not be wrong or unprecedented; for why might we not in our own right (which is the same right that everyone has) criticise or condemn anything which they have observed or assumed wrongly? And yet however justified or legitimate, the contest itself would still have been unequal because of the limitations of our resources. But since our concern is to open up a completely different way to the intellect, unknown and untried by the ancients, the situation is quite different; parties and partisanship are out; our role is merely that of a guide, and this surely carries little authority, and depends on fortune rather than on ability and excellence. And this kind of remark applies to persons; the following one to things themselves.

We have no intention of dethroning the prevailing philosophy, or any other now or in the future that may be more correct or complete. Nor do we want to stop this accepted philosophy and others of its kind from fuelling disputations, adorning discourses and being successfully employed in academic instruction and handbooks of civil life. In fact we frankly admit and declare that the philosophy which we are introducing will be quite useless for those purposes. It is not easy to get hold of, it cannot be picked up in passing, it does not flatter intellectual prejudices, it will not adapt itself to the common understanding except in its utility and effects.

Let there be two sources of learning therefore, and two means of dissemination (and may this be good and fortunate for both of them). Let there likewise be two clans or families of thinkers or philosophers; and let them not be hostile or alienated from each other, but allies bound together by ties of mutual assistance. And above all let there be one method for cultivating the sciences and a different method for discovering them. Those to whom the first method is preferable and more acceptable, whether because of their haste or for reasons of civil life, or because they lack the intellectual capacity to grasp and master the other method, we pray that their activities go well for them and as they desire, and that they get what they are after. But any man whose care and concern is not merely to be content with what has been discovered and make use of it, but to penetrate further; and not to defeat an opponent in argument but to conquer nature by action; and not to have nice, plausible opinions about things but sure, demonstrable knowledge; let such men (if they please), as true sons of the sciences, join with me, so that we may pass the antechambers of nature which innumerable others have trod, and eventually open up access to the inner rooms. For better understanding, and to make what we mean more familiar by assigning names, we have chosen to call the one way or method the *Anticipation of the Mind*[22] and the other the *Interpretation of Nature.*

There is also a request which it seems we must make. We have thought hard and taken care that our proposals should not only be true but should enter men's minds easily and smoothly (occupied and blocked as they are in different ways). But it is reasonable for us to request (especially in such a renewal of learning and the sciences) that no one who wishes to judge or reflect upon these our thoughts, whether of his own sense or with a host of authorities or by the forms of demonstration (which have the authority at present of judicial rules), should expect to be able to do this casually or while he is about something else, but should get to know the subject properly; should himself try a little the road which we are designing and building; should get used to the subtlety of things which experience suggests; should finally correct, within a fair and reasonable time, the bad mental habits which are so deeply ingrained; and then and only then (if he so

[22] On 'anticipation of nature' see I.26ff.

please), after he has grown up and become his own master, let him use his own judgement.

THERE FOLLOWS
THE SUMMARY OF THE SECOND PART
DIGESTED INTO
APHORISMS

SUMMARY
OF THE SECOND PART,
DIGESTED
INTO
APHORISMS

APHORISMS
ON THE INTERPRETATION OF NATURE
AND ON THE KINGDOM OF MAN
[BOOK I]

Aphorism I

Man is Nature's agent and interpreter; he does and understands only as much as he has observed of the order of nature in fact or by inference; he does not know and cannot do more.

II

Neither the bare hand nor the unaided intellect has much power; the work is done by tools and assistance, and the intellect needs them as much as the hand. As the hand's tools either prompt or guide its motions, so the mind's tools either prompt or warn the intellect.

III

Human knowledge and human power come to the same thing, because ignorance of cause frustrates effect. For Nature is conquered only by obedience; and that which in thought is a cause, is like a rule in practice.

IV

All man can do to achieve results[1] is to bring natural bodies together and take them apart; Nature does the rest internally.

---

[1] *opus, opera*: widely used throughout *The New Organon* in a variety of senses, this word has been translated, according to context, as 'results' or 'effects' or 'work'. The related *operatio* is usually given as 'operation' or 'practice', and *operativus* as 'practical' or 'applied' (as in *pars operativa*).

V

Mechanic, mathematician, physician, alchemist and magician do meddle with nature (for results); but all, as things are, to little effect and with slender success.

VI

There is something insane and self-contradictory in supposing that things that have never yet been done can be done except by means never tried.

VII

The creations of mind and hand look quite prolific in books and manufactures. But all that varied production consists in extreme subtlety and in deductions from a few things that have become known, not in the number of Axioms.

VIII

Even the results which have been discovered already are due more to chance and experience than to sciences; for the sciences we now have are no more than elegant arrangements of things previously discovered, not methods of discovery or pointers to new results.

IX

The cause and root of nearly all the deficiencies of the sciences is just this: that while we mistakenly admire and praise the powers of the human mind, we do not seek its true supports.

X

The subtlety of nature far surpasses the subtlety of sense and intellect, so that men's fine[2] meditations, speculations and endless discussions are quite insane, except that there is no one who notices.

---

[2] *Pulcher* seems always to be used ironically in *The New Organon*.

XI

As the sciences in their present state are useless for the discovery of works, so logic in its present state is useless for the discovery of sciences.

XII

Current logic is good for establishing and fixing errors (which are themselves based on common notions) rather than for inquiring into truth; hence it is not useful, it is positively harmful.

XIII

The syllogism is not applied to the principles of the sciences, and is applied in vain to the middle axioms, since it is by no means equal to the subtlety of nature. It therefore compels assent without reference to things.[3]

XIV

The syllogism consists of propositions, propositions consist of words, and words are counters for notions. Hence if the notions themselves (this is the basis of the matter) are confused and abstracted from things without care, there is nothing sound in what is built on them. The only hope is true *induction*.

XV

There is nothing sound in the notions of logic and physics: neither *substance*, nor *quality*, nor *action* and *passion*, nor *being* itself are good notions; much less *heavy, light, dense, rare, wet, dry, generation, corruption, attraction, repulsion, element, matter, form* and so on; all fanciful and ill defined.[4]

XVI

The notions of the lowest species, *man, dog, dove*, and of the immediate perceptions of sense, *hot, cold, white, black*, do not much mislead, though, from

---

3 Cf. 1.29.

4 I owe 'all fanciful and ill defined' to Kitchin's translation.

the flux of matter and the conflict of things, they are sometimes confused; all the others (that men have so far made use of) are aberrations, not being drawn and abstracted from things in the proper ways.

XVII

Passion and aberration occur no less in forming axioms than in abstracting notions, even in the principles that depend on ordinary induction. But this is much more the case with the lower axioms and propositions which the syllogism generates.

XVIII

The things that have hitherto been discovered in the sciences all fit nicely into common notions; in order to penetrate to the more inward and remote parts of nature, both notions and axioms must be abstracted from things in a more certain, better-grounded way; and a more certain and altogether better intellectual procedure must come into use.

XIX

There are, and can be, only two ways to investigate and discover truth. The one leaps from sense and particulars to the most general axioms, and from these principles and their settled truth, determines and discovers intermediate axioms; this is the current way. The other elicits axioms from sense and particulars, rising in a gradual and unbroken ascent to arrive at last at the most general axioms; this is the true way, but it has not been tried.

XX

Left to itself the intellect goes the same way as it does when it follows the order of dialectic (i.e. the first of the two ways above). The mind loves to leap to generalities, so that it can rest; it only takes it a little while to get tired of experience. These faults have simply been magnified by dialectic, for ostentatious disputes.

XXI

In a sober, grave and patient character the intellect left to itself (especially if unimpeded by received doctrines) makes some attempt on that other way, which is the right way, but with little success; since without guidance and assistance it is a thing inadequate and altogether incompetent to overcome the obscurity of things.

XXII

Both ways start from sense and particulars, and come to rest in the most general; but they are vastly different. For one merely brushes experience and particulars in passing, the other deals fully and properly with them; one forms certain abstract and useless generalities from the beginning, the other rises step by step to what is truly better known by nature.

XXIII

There is a great distance between the illusions[5] of the human mind and the ideas of the divine mind; that is, between what are no more than empty opinions and what we discover are the true prints and signatures made on the creation.

XXIV

Axioms formed by argumentation cannot be good at all for the discovery of new results, because the subtlety of nature far surpasses the subtlety of argumentation. But arguments duly and properly abstracted from particulars readily indicate and suggest new particulars; this is what makes the sciences practical.

XXV

The axioms currently in use have come from limited and common experience and the few particulars that occur most often, and are more or less made and stretched to fit them; so that it is no surprise if they do not lead

[5] *idola*. See further note on I.39.

to new particulars. And if it happens that a new instance turns up which was not previously noticed or known, the axiom is saved by some frivolous distinction when it would be more truthful to amend the axiom.

XXVI

For the purposes of teaching, we have chosen to call the reasoning which men usually apply to nature *anticipations of nature* (because it is a risky, hasty business), and to call the reasoning which is elicited from things in proper ways *interpretation of nature*.

XXVII

Anticipations are quite strong enough to induce agreement, since even if men were mad in one common way together, they could agree among themselves well enough.

XXVIII

In fact *anticipations* are much more powerful in winning assent than *interpretations*; they are gathered from just a few instances, especially those which are common and familiar, which merely brush past the intellect and fill the imagination. *Interpretations* by contrast are gathered piece by piece from things which are quite various and widely scattered, and cannot suddenly strike the intellect. So that, to common opinion, they cannot help seeming hard and incongruous, almost like mysteries of faith.

XXIX

In sciences which are based on opinions and accepted views, the use of *anticipations* and dialectic is acceptable where the need is to compel assent without reference to things.

XXX

Even if all the minds of all the ages should come together and pool their labours and communicate their thoughts, there will be no great progress made in sciences by means of *anticipations*, because errors which are radical and lie in the fundamental organisation of the mind are not put right by subsequent efforts and remedies, however brilliant.

XXXI

It is futile to expect a great advancement in the sciences from overlaying[6] and implanting new things on the old; a new beginning[7] has to be made from the lowest foundations, unless one is content to go round in circles for ever, with meagre, almost negligible, progress.

XXXII

The honour of the ancient authors stands firm, and so does everyone's honour; we are not introducing a comparison of minds or talents but a comparison of ways; and we are not taking the role of a judge but of a guide.

XXXIII

No judgement can rightly be made of our way (one must say frankly), nor of the discoveries made by it, by means of *anticipations* (i.e. the reasoning currently in use); for one must not require it to be approved by the judgement of the very thing which is itself being judged.

XXXIV

There is no easy way of teaching or explaining what we are introducing; because anything new will still be understood by analogy with the old.

XXXV

Borgia said of the expedition of the French into Italy that they had come with chalk in their hands to mark their billets, not with arms to force their way through. Likewise it is our design for our teaching to find its way into suitable, capable minds; there is no room for refutations when we disagree about principles and notions themselves and even about the forms of proof.

---

[6] *superinduco*, normally translated 'superinduce': cf. II.1 etc.
[7] *instauratio*, normally translated 'renewal', as in the title of the whole work, *The Great Renewal*.

XXXVI

There remains one simple way of getting our teaching across, namely to introduce men to actual particulars and their sequences and orders, and for men in their turn to pledge to abstain for a while from notions, and begin to get used to actual things.

XXXVII

In its initial positions our way agrees to some extent with the method of the supporters of *lack of conviction*;[8] but in the end our ways are far apart and strongly opposed. They assert simply that nothing can be known; but we say that not much can be known in nature by the way which is now in use. They thereupon proceed to destroy the authority of sense and intellect; but we devise and provide assistance to them.

XXXVIII

The *illusions* and false notions which have got a hold on men's intellects in the past and are now profoundly rooted in them, not only block their minds so that it is difficult for truth to gain access, but even when access has been granted and allowed, they will once again, in the very renewal of the sciences, offer resistance and do mischief unless men are forewarned and arm themselves against them as much as possible.

XXXIX

There are four kinds of *illusions* which block men's minds. For instruction's sake, we have given them the following names: the first kind are called *idols of the tribe*; the second *idols of the cave*; the third *idols of the marketplace*; the fourth *idols of the theatre*.[9]

---

[8] Bacon uses the Greek term *acatalepsia*, the mark of ancient Sceptics. At 1.61 he distinguishes them from what he regards as Pyrrho's more radical school of ancient Sceptics, whom he calls *Ephectici*, 'those who suspend judgement'.
[9] It has seemed better to accept the time-honoured translations of the four classes of *idola*, as given in this aphorism. We have however sometimes chosen to translate *idola* itself, apart from these four phrases, as 'illusions'.

XL

Formation of notions and axioms by means of true *induction* is certainly an appropriate way to banish *idols* and get rid of them; but it is also very useful to identify the *idols*. Instruction about *idols* has the same relation to the *interpretation of nature* as teaching the *sophistic refutations* has to ordinary logic.

XLI

The *idols of the tribe* are founded in human nature itself and in the very tribe or race of mankind. The assertion that the human senses are the measure of things is false; to the contrary, all perceptions, both of sense and mind, are relative to man, not to the universe. The human understanding is like an uneven mirror receiving rays from things and merging its own nature with the nature of things, which thus distorts and corrupts it.

XLII

The *idols of the cave* are the illusions of the individual man. For (apart from the aberrations of human nature in general) each man has a kind of individual cave or cavern which fragments and distorts the light of nature. This may happen either because of the unique and particular nature of each man; or because of his upbringing and the company he keeps; or because of his reading of books and the authority of those whom he respects and admires; or because of the different impressions things make on different minds, preoccupied and prejudiced perhaps, or calm and detached, and so on. The evident consequence is that the human spirit (in its different dispositions in different men) is a variable thing, quite irregular, almost haphazard. Heraclitus[10] well said that men seek knowledge in lesser, private worlds, not in the great or common world.

XLIII

There are also *illusions* which seem to arise by agreement and from men's association with each other, which we call *idols of the marketplace;* we take

---

[10] Heraclitus, fr. 2. A Greek philosopher of the late sixth century, Heraclitus of Ephesus is one of the early Greek philosophers whom as a group Bacon much admired. See 1.68.

the name from human exchange and community. Men associate through talk; and words are chosen to suit the understanding of the common people. And thus a poor and unskilful code of words incredibly obstructs the understanding. The definitions and explanations with which learned men have been accustomed to protect and in some way liberate themselves, do not restore the situation at all. Plainly words do violence to the understanding, and confuse everything; and betray men into countless empty disputes and fictions.

XLIV

Finally there are the *illusions* which have made their homes in men's minds from the various dogmas of different philosophies, and even from mistaken rules of demonstration. These I call *idols of the theatre*, for all the philosophies that men have learned or devised are, in our opinion, so many plays produced and performed which have created false and fictitious worlds. We are not speaking only of the philosophies and sects now in vogue or even of the ancient ones; many other such plays could be composed and concocted, seeing that the causes of their very different errors have a great deal in common. And we do not mean this only of the universal philosophies, but also of many principles and axioms of the sciences which have grown strong from tradition, belief and inertia. But we must speak at greater length and separately of each different kind of *idol*, to give warning to the human understanding.

XLV

The human understanding from its own peculiar nature willingly supposes a greater order and regularity in things than it finds, and though there are many things in nature which are unique and full of disparities, it invents parallels and correspondences and non-existent connections. Hence those false notions that *in the heavens all things move in perfect circles* and the total rejection of spiral lines and dragons (except in name). Hence the element of fire and its orbit have been introduced to make a quaternion with the other three elements, which are accessible to the senses. Also a ratio of ten to one is arbitrarily imposed on the elements (as they call them), which is the ratio of their respective rarities; and other such nonsense. This vanity prevails not only in dogmas but also in simple notions.

XLVI

Once a man's understanding has settled on something (either because it is an accepted belief or because it pleases him), it draws everything else also to support and agree with it. And if it encounters a larger number of more powerful countervailing examples, it either fails to notice them, or disregards them, or makes fine distinctions to dismiss and reject them, and all this with much dangerous prejudice, to preserve the authority of its first conceptions. So when someone was shown a votive tablet in a temple dedicated, in fulfilment of a vow, by some men who had escaped the danger of shipwreck, and was pressed to say whether he would now recognise the divinity of the gods, he made a good reply when he retorted: 'Where are the offerings of those who made vows and perished?'[11] The same method is found perhaps in every superstition, like astrology, dreams, omens, divine judgements and so on: people who take pleasure in such vanities notice the results when they are fulfilled, but ignore and overlook them when they fail, though they do fail more often than not. This failing finds its way into the sciences and philosophies in a much more subtle way, in that once something has been settled, it infects everything else (even things that are much more certain and powerful), and brings them under its control. And even apart from the pleasure and vanity we mentioned, it is an innate and constant mistake in the human understanding to be much more moved and excited by affirmatives than by negatives, when rightly and properly it should make itself equally open to both; and in fact, to the contrary, in the formation of any true axiom, there is superior force in a negative instance.

XLVII

The human understanding is most affected by things which have the ability to strike and enter the mind all at once and suddenly, and to fill and expand the imagination. It pretends and supposes that in some admittedly imperceptible way, everything else is just like the few things that took the mind by storm. The understanding is very slow and ill adapted to make the long journey to those remote and heterogeneous instances which test axioms as in a fire, unless it is made to do so by harsh rules and the force of authority.

[11] This story is told of Diagoras the Atheist at Cicero, *On the Nature of the Gods*, III.37, and of Diogenes the Cynic at Diogenes Laertius, *Lives of Eminent Philosophers*, VI.59.

XLVIII

The human understanding is ceaselessly active, and cannot stop or rest, and seeks to go further; but in vain. Therefore it is unthinkable that there is some boundary or farthest point of the world; it always appears, almost by necessity, that there is something beyond. Again it cannot be conceived how eternity has come down to this day; since the distinction which is commonly accepted that there is *an infinity of the past and an infinity of the future* can no way stand, because it would follow that there is one infinity which is greater than another infinity, and that infinity is being consumed and tends towards the finite. There is a similar subtlety about ever divisible lines, from thought's lack of restraint. This indiscipline of the mind works with greater damage on the discovery of causes: for though the most universal things in nature must be brute facts,[12] which are just as they are found, and are not themselves truly causable, the human understanding, not knowing how to rest, still seeks things better known.[13] And then as it strives to go further, it falls back on things that are more familiar, namely final causes, which are plainly derived from the nature of man rather than of the universe, and from this origin have wonderfully corrupted philosophy. It is as much a mark of an inept and superficial thinker to look for a cause in the most universal cases as not to feel the need of a cause in subordinate and derivative cases.

XLIX

The human understanding is not composed of dry light,[14] but is subject to influence from the will and the emotions, a fact that creates fanciful knowledge; man prefers to believe what he wants to be true. He rejects what is difficult because he is too impatient to make the investigation; he rejects sensible ideas, because they limit his hopes; he rejects the deeper truths of nature because of superstition; he rejects the light of experience, because he is arrogant and fastidious, believing that the mind should not be seen to be spending its time on mean, unstable things; and he rejects anything unorthodox because of common opinion. In short, emotion marks and

---

[12] *positiva*: cf. II.48 (14) on some things which 'should be accepted on the basis of experience and as brute facts'.

[13] *notiora*, perhaps equivalent to *natura notiora*, 'better known in nature'.

[14] Cf. Heraclitus, fr. 118.

stains the understanding in countless ways which are sometimes impossible to perceive.

L

But much the greatest obstacle and distortion of human understanding comes from the dullness, limitations and deceptions of the senses; so that things that strike the senses have greater influence than even powerful things which do not directly strike the senses. And therefore thought virtually stops at sight; so that there is little or no notice taken of things that cannot be seen. And so all operation of spirits enclosed in tangible bodies remains hidden and escapes men's notice. And all the more subtle structural change[15] in the parts of dense objects (which is commonly called alteration, although in truth it is movement of particles) is similarly hidden. Yet unless the two things mentioned are investigated and brought to the light, nothing important can be done in nature as far as results are concerned. Again, the very nature of the common air and of all the bodies which surpass air in rarity (of which there are many) is virtually unknown. For by itself sense is weak and prone to error, nor do instruments for amplifying and sharpening the senses do very much. And yet every interpretation of nature which has a chance to be true is achieved by instances, and suitable and relevant experiments, in which sense only gives a judgement on the experiment, while the experiment gives a judgement on nature and the thing itself.

LI

The human understanding is carried away to abstractions by its own nature, and pretends that things which are in flux are unchanging. But it is better to dissect nature than to abstract; as the school of Democritus[16] did, which penetrated more deeply into nature than the others. We should study matter, and its structure (*schematismus*), and structural change (*meta-schematismus*), and pure act, and the law of act or motion; for *forms* are figments of the human mind, unless one chooses to give the name of *forms* to these laws of act.

---

[15] *meta-schematismus*
[16] Democritus of Abdera, Greek atomist philosopher of the fifth century BC.

LII

Such then are the *illusions* that we call *idols of the tribe*, which have their origin either in the regularity of the substance of the human spirit; or in its prejudices; or in its limitations; or in its restless movement; or in the influence of the emotions; or in the limited powers of the senses; or in the mode of impression.

LIII

Idols of the cave have their origin in the individual nature of each man's mind and body; and also in his education, way of life and chance events. This category is varied and complex, and we shall enumerate the cases in which there is the greatest danger and which do most to spoil the clarity of the understanding.

LIV

Men fall in love with particular pieces of knowledge and thoughts: either because they believe themselves to be their authors and inventors; or because they have put a great deal of labour into them, and have got very used to them. If such men betake themselves to philosophy and universal speculation, they distort and corrupt them to suit their prior fancies. This is seen most conspicuously in Aristotle,[17] who utterly enslaved his natural philosophy to his logic, and made it a matter of disputation and almost useless. Chemists as a group have built up a fantastic philosophy out of a few experiments at the furnace, which has a very limited range; and Gilbert[18] too, after his strenuous researches on the magnet, immediately concocted a philosophy in conformity with the thing that had the dominating influence over him.

LV

The biggest, and radical, difference between minds as far as philosophy and the sciences is concerned, is this: that some minds are more effective

---

[17] The great Greek philosopher, 384-322 BC, Bacon's particular target in *The New Organon*. See 1.63 and 67, and *Introduction*, pp. 7–8.

[18] William Gilbert (1544–1603), scientist and physician. Court physician to Elizabeth I and James I. Best known for his studies of magnetism, he published *De magnete* in 1600.

and more suited to noticing the differences between things, others to noticing their similarities. For sharp and steady minds can fix their attention, and concentrate for long periods on every subtle difference; but sublime and discursive minds discern even the slightest and most general similarities in things, and bring them into relationship; both minds easily go to extremes by grasping at degrees of things or at shadows.

LVI

There are some minds which are devoted to admiration for antiquity, others to the love and embrace of novelty, and few have the temperament to keep to the mean without criticising the true achievements of the ancients or despising the real contributions of the moderns. This is a great loss to the sciences and to philosophy, since these are not judgements but enthusiasm for antiquity or modernity; and truth is not to be sought from the felicity of a particular time, which is a variable thing, but from the light of nature, which is eternal. We must reject these enthusiasms, and ensure that the understanding is not diverted into compliance with them.

LVII

Observation of nature and of bodies in their simple parts fractures and diminishes the understanding; observation of nature and of bodies in their composition and complex structure stupefies and confounds the understanding. This is best seen in a comparison of the school of Leucippus and Democritus[19] with the other philosophies. It is so concerned with the particles of things that it almost forgets their structures; while the others are so astonished by beholding the structures that they do not penetrate to the simple parts of nature. These kinds of observation therefore need to be alternated and taken in turn, so that the understanding may be rendered both penetrating and comprehensive; and the defects we mentioned avoided, with the illusions they generate.

LVIII

Let such be the care in observation which will banish and get rid of *idols of the cave*, which mostly have their origin in a dominance or excess of

---

[19] Greek atomists. See n. 16 above.

composition and division, or in partiality for historical periods, or in the large and minute objects. And in general every student of nature must hold in suspicion whatever most captures and holds his understanding; and this warning needs to be all the more applied in issues of this kind, to keep the understanding clear and balanced.

LIX

But the *idols of the marketplace*[20] are the biggest nuisance of all, because they have stolen into the understanding from the covenant[21] on words and names. For men believe that their reason controls words. But it is also true that words retort and turn their force back upon the understanding; and this has rendered philosophy and the sciences sophistic and unproductive. And words are mostly bestowed to suit the capacity of the common man, and they dissect things along the lines most obvious to the common understanding. And when a sharper understanding, or more careful observation, attempts to draw those lines more in accordance with nature, words resist. Hence it happens that the great and solemn controversies of learned men often end in disputes about words and names. But it would be wiser (in the prudent manner of the mathematicians) to begin with them, and to reduce them to order by means of definitions. However, in the things of nature and matter, these definitions cannot cure this fault. For the definitions themselves consist of words, and words beget words, so that it is necessary to have recourse to particular instances and their sequences and orders; as we shall explain soon when we deal with the method and manner of forming notions and axioms.

LX

The *illusions* which are imposed on the understanding by words are of two kinds. They are either names of things that do not exist (for as there are things that lack names because they have not been observed, so there are also names that lack things because they have been imaginatively assumed), or they are the names of things which exist but are

---

[20] We have retained the traditional translation of this phrase because it is so familiar. However, 'marketplace' has quite the wrong connotation in suggesting some economic notion. A better rendering of 'forum' would be 'townsquare'; it is the place where men meet and talk and reinforce each others' 'idols', or 'illusions'. On 'idols' see 'Plan of the Work', n. 13.

[21] *foedus*: for agreement as the origin of the meaning of words, see above, I.43.

confused and badly defined, being abstracted from things rashly and unevenly. Of the former sort are fortune, the first mover, the orbs of the planets, the element of fire and fictions of that kind, which owe their origin to false and groundless theories. *Idols* of this kind are easily got rid of; they can be eradicated by constantly rejecting and outdating the theories.

But the other kind of *idol* is complex and deep-seated, being caused by poor and unskilful abstraction. For example, let us take a word ('wet' if you like) and see how the things signified by this word go together; it will be found that the word 'wet' is simply an undiscriminating token for different actions which have no constancy or common denominator. For it signifies both what is easily poured around another object; and what is without its own boundaries and unstable; and what easily gives way all round; and what easily divides and disperses; and what easily combines and comes together; and what easily flows and is set in motion; and what easily adheres to another body and makes it wet; and what is easily reduced to a liquid, or liquefies, from a previous solid state. Hence when it comes to predicating and applying this word, if you take it one way, a flame is wet; if in another, air is not wet; if in another, a speck of dust is wet; if in another, glass is wet; it is easily seen that this notion has been rashly abstracted from water and common and ordinary liquids only, without any proper verification.

There are various degrees of deficiency and error in words. Least faulty is the class of names of particular substances, especially the lowest, well-derived species (e.g. the notions of chalk and mud are good, of earth bad); next is the class of names of actions, such as 'generate', 'corrupt', 'alter'; the faultiest class is of the names of qualities (with the exception of direct objects of sense), such as 'heavy', 'light', 'rare', 'dense' etc.; but in all classes, inevitably, some notions are a little better than others, depending on the number of each that come to the notice of the human senses.

LXI

Idols of the theatre are not innate or stealthily slipped into the understanding; they are openly introduced and accepted on the basis of fairytale theories and mistaken rules of proof. It is not at all consistent with our argument to attempt or undertake to refute them.

There is no possibility of argument, since we do not agree either about the principles or about the proofs.[22] It is a happy consequence that the ancients may keep their reputation. I take nothing from them, since the question is simply about the way. As the saying goes, a lame man on the right road beats the runner who misses his way. It is absolutely clear that if you run the wrong way, the better and faster you are, the more you go astray.

Our method of discovery in the sciences is designed not to leave much to the sharpness and strength of the individual talent; it more or less equalises talents and intellects. In drawing a straight line or a perfect circle, a good deal depends on the steadiness and practice of the hand, but little or nothing if a ruler or a compass is used. Our method is exactly the same. But though there is no point in specific refutations, something must be said about the sects and kinds of such theories; and then of the external signs that the situation is bad; and lastly of the reasons for so much failure, and such persistent and general agreement in error; so that there may be easier access to true things, and the human understanding may be more willing to cleanse itself and dismiss its *idols*.

LXII

There are many *idols of the theatre*, or theories, and there could be many more, and perhaps one day there will be. For if men's minds had not been preoccupied for so many centuries now with religion and theology, and if also civil governments (especially monarchies) had not been hostile to such novelties even in thought, so that men could not get involved in them without danger and damage to their fortunes, and would not only be deprived of reward but exposed to contempt and envy, without doubt a number of other philosophical and theoretical sects would have been introduced, like those which once flourished in great variety in ancient Greece. For just as several accounts of the heavens can be fashioned from the *phenomena* of the air, so, and much more, various dogmas can be based and constructed upon the phenomena of philosophy. And the stories of this kind of *theatre* have something else in common with the dramatist's theatre, that narratives made up for the stage are neater and more elegant than true stories from history, and are the sort of thing people prefer.

In general, for the content of philosophy, either much is made of little or

---

22 This is a legal maxim which Bacon adapts to his method.

little is made of much, so that in both cases philosophy is built upon an excessively narrow basis of experience and natural history, and bases its statements on fewer instances than is proper. Philosophers of the rational type are diverted from experience by the variety of common phenomena, which have not been certainly understood or carefully examined and considered; they depend for the rest on reflection and intellectual exercise.

There are also philosophers of another type who have laboured carefully and faithfully over a few experiments, and have had the temerity to tease out their philosophies from them and build them up; the rest they twist to fit that pattern in wonderful ways.

There is also a third type, who from faith and respect mingle theology and traditions; some of them have been unfortunately misled by vanity to try to derive sciences from Spirits and Genii. And so the root of errors and false philosophy is of three kinds: Sophistic, Empirical and Superstitious.

LXIII

The most obvious example of the first type is Aristotle, who spoils natural philosophy with his dialectic. He constructed the world of categories; he attributed to the human soul the noblest substance, a genus based on words of second intention; he transformed the interaction of *dense* and *rare*, by which bodies occupy greater and smaller dimensions or spaces, into the unilluminating distinction between act and potentiality; he insisted that each individual body has a unique and specific motion, and if they participate in some other motion, that motion is due to a different reason; and he imposed innumerable other things on nature at his own whim. He was always more concerned with how one might explain oneself in replying, and to giving some positive response in words, than of the internal truth of things; and this shows up best if we compare his philosophy with other philosophies in repute among the Greeks. The 'similar substances'[23] of Anaxagoras, the atoms of Leucippus and Democritus, the earth and sky of Parmenides, the strife and friendship of Empedocles, the dissolution of bodies into the undifferentiated nature of fire and their return to solidity in Heraclitus, all have something of natural philosophy in them, and have the feel of nature and experience and bodies;[24] whereas Aristotle's physics too often sound like mere terms of dialectic, which he rehashed under a

---

[23] *homoiomera*
[24] These are all pre-Socratic Greek philosophers of the late sixth and the fifth centuries BC.

more solemn name in his metaphysics, claiming to be more of a realist, not a nominalist. And no one should be impressed because in his books *On Animals* and in his *Problems* and other treatises there is often discussion of experiments. He had in fact made up his mind beforehand, and did not properly consult experience as the basis of his decisions and axioms; after making his decisions arbitrarily, he parades experience around, distorted to suit his opinions, a captive. Hence on this ground too he is guiltier than his modern followers (the scholastic philosophers) who have wholly abandoned experience.

LXIV

The *empirical* brand of philosophy generates more deformed and freakish dogmas than the *sophistic* or rational kind, because it is not founded on the light of common notions (which though weak and superficial, is somehow universal and relevant to many things) but on the narrow and unilluminating basis of a handful of experiments. Such a philosophy seems probable and almost certain to those who are engaged every day in experiments of this kind and have corrupted their imagination with them; to others it seems unbelievable and empty. There is a notable example of this among the chemists and their dogmas; otherwise it scarcely exists at this time, except perhaps in the philosophy of Gilbert. However, we should not fail to give a warning about such philosophies. We already conceive and foresee that, if ever men take heed of our advice and seriously devote themselves to experience (having said goodbye to the sophistic doctrines), then this philosophy will at last be genuinely dangerous, because of the mind's premature and precipitate haste, and its leaping or flying to general statements and the principles of things; even now we should be facing this problem.

LXV

The corruption of philosophy from *superstition* and a dash of theology is much more widely evident, and causes a very great deal of harm either to entire philosophies or to their parts. For the human mind is no less liable to the impressions of fantasy than to impressions from common notions. The disputatious and *sophistical* kind of philosophy catches the understanding in a trap, but the other kind, the fantastic, high-blown, semi-poetical philosophy, seduces it. There is in man a kind of ambition of the intellect no less than of the will, especially in lofty, high-minded characters.

A conspicuous example of this occurs among the Greeks in Pythagoras, where it is combined with a rather crass and cumbrous *superstition*, and in a more perilous and subtle form in Plato and his school. This kind of evil also occurs in parts of other philosophies by the introduction of abstract forms and final causes and first causes, and by frequent omission of inter- mediate causes and so on. We must give the strongest warning here. For the worst thing is the *apotheosis* of error; respect for foolish notions has to be regarded as a disease of the intellect. Some of the moderns have, with extreme frivolity, been so lenient to such foolishness that they have tried to base natural philosophy on Genesis and the Book of Job and other sacred Scriptures, *seeking the dead among the living*.[25] This folly needs to be checked and stifled all the more vigorously because heretical religion as well as fanciful philosophy derives from the unhealthy mingling of divine and human. And therefore it is very salutary, in all sobriety, to give to faith only what belongs to faith.

LXVI

So much for the poor authority of philosophies founded on *common notions* or *few experiments* or *superstition*. We must next speak of poor material for reflection, especially in natural philosophy. The human mind is misled by looking at what is done in the mechanical arts, in which bodies are entirely changed by composition and separation, into supposing that something similar also happens in the universal nature of things. This is the source of the fiction of the *elements* and of their *collision* to form natural bodies. Again, when a man contemplates the liberty of nature, he comes upon the species of things - animals, plants and minerals; and from there he slides easily into the thought that there are in nature certain primary forms of things, which nature is struggling to bring out, and that all the rest of her variety comes from the obstacles and errors of nature in completing her task, or from conflict between the different species. The first thought gave us the first elementary qualities, the second occult properties and specific virtues; both notions belong to the category of meaningless summaries of

[25] Luke 24:5.

observations, in which the mind acquiesces and is diverted from sounder ideas. Physicians have more success when they use the secondary qualities of things and the operations of attraction, repulsion, rarefaction, condensation, dilation, contraction, dissipation, maturation and so on. They would have made even better progress if from the summary notions I have spoken of (i.e. elementary qualities and specific virtues) they did not corrupt the others (which have been rightly noted), by reducing them to first qualities and subtle, incommensurable mixtures of things, or by not extending them to third and fourth qualities by further, more careful observation; they broke off their observations too early. We should look for such virtues (I do not say the same but similar) not only in the medicines for the human body, but also in the factors which modify other natural bodies.

It is a much more serious problem that they observe and investigate the principles of things at rest *from which* things come into being, and not the moving things *through which* things come into being. The former relate to discussion, the latter to results. And there is no value in the usual differences of motion noted in traditional natural philosophy – *generation, corruption, growth, decrease, alteration and movement*. For this is all they mean: if a body, otherwise unchanged, moves spatially, this is *movement*;[26] if while place and species remain the same, it changes in quality, this is *alteration*; and if as a result of that change, the mass itself and quantity of the body do not remain the same, this is the motion of *growth* and *decrease*; if they alter to the extent that they change the very species and substance, and become other things, that is *generation* and *corruption*. But these are merely popular notions, and do not pierce into the nature at all; and they are measures and periods only, not species of motion. For they indicate only *how far?* and not *how*, or *from what source*. They tell us nothing of the appetite of bodies, or of the process of their parts; they merely guess at a division when a motion shows the senses in an obvious manner that an object is otherwise than it was before. And when they want to explain something of the causes of motions, and to set up a division of them, they introduce a distinction between natural and violent motion, a supremely idle move, since this distinction comes straight from the common notion. For every violent motion is in truth also a natural motion, i.e. when an external cause sets a nature in motion in a different manner than it was in before.

---

[26] *latio*

But leave all this aside; if anyone (for the sake of example) has observed that there is in bodies an appetite for mutual contact, so that they do not suffer the unity of nature to be completely pulled and torn apart, and thereby have a vacuum; if anyone has observed that there is in bodies an appetite for withdrawing to their own natural size or tension, so that if they are compressed or stretched more or less than this, they instantly strive to recover and retrieve their former sphere and extension; or if anyone has observed that there is in bodies an appetite to assemble with masses of things of their own kind, i.e. an appetite in dense things for the earth, of thin and rare things for the circuit of the sky: these things and things like them are truly physical kinds of motion. But those others are simply theoretical and scholastic, as is manifestly clear from this comparison between them.

It is no less of a problem that in their philosophies and observations they waste their efforts on investigating and treating the principles of things and the ultimate causes of nature (*ultimatibus naturae*), since all utility and opportunity for application lies in the intermediate causes (*in mediis*). This is why men do not cease to abstract nature until they reach potential and unformed matter, nor again do they cease to dissect nature till they come to the atom. Even if these things were true, they can do little to improve men's fortunes.

LXVII

The understanding needs also to be cautioned against the intemperance of philosophies in giving or refusing assent; such intemperance seems to fix *idols* and somehow prolong their life, so that there is no possibility of getting rid of them.

There are two kinds of excess: one is that of those who readily *make pro-nouncements*, and make the sciences lay down the law in a magisterial manner; the other is the excess of those who have introduced *lack of conviction* (*acatalepsia*) and an aimless and endless questioning. The first of these represses the understanding, the second robs it of strength. For after the philosophy of Aristotle had (in the manner of the Ottomans towards their brothers) slaughtered the other philosophies with vicious disputations, it made pronouncements on every single question; and he himself formulates objections at his own whim, and then deals with them, so that everything is certain and settled; and this is still the way with his successors.

The school of Plato introduced *lack of conviction*, at first apparently in jest and irony, from resentment against the old sophists, Protagoras, Hippias and the others, who feared nothing so much as to be seen to be hesitating about anything.[27] The New Academy made *lack of conviction* a dogma, and held it as a tenet.[28] It is a more honest method than a licence to *make pronouncements*, since they claim for themselves that they do not subvert inquiry, as Pyrrho and the Ephectici[29] did, but have something to follow as probable, though nothing to hold as true. Nevertheless, after the human mind has once despaired of finding truth, everything becomes very much feebler; and the result is that they turn men aside to agreeable discussions and discourses, and a kind of ambling around things, rather than sustain them in the severe path of inquiry. As we have said from the beginning, and constantly maintain, we must not detract from the authority of the human senses and the human understanding and their deficiencies, but must find assistance for them.

LXVIII

So much for the individual kinds of *idols* and their trappings; all of which must be rejected and renounced and the mind totally liberated and cleansed of them, so that there will be only one entrance into the kingdom of man, which is based upon the sciences, as there is into the kingdom of heaven, 'into which, except as an infant, there is no way to enter'.[30]

LXIX

Bad demonstrations are the defences and fortresses of the *idols*; and the demonstrations which we have in dialectic do no more than addict and enslave the world wholly to human thoughts and thoughts to words. Demonstrations are, potentially, philosophies and sciences themselves. For as they are, and as they are well or badly devised, philosophies and

---

27 This refers to the sceptical side of the Platonic legacy, which arose from the questioning dialectic of Socrates. Protagoras and Hippias are two of the best-known names of the fifth-century Sophistic movement; they prided themselves, as teachers of rhetoric, on their ability to speak eloquently and positively on any question.

28 The New Academy is the name given to the academy founded by Plato, when the Sceptic Carneades became head in the mid second century BC.

29 *Ephectici*, 'those who suspend judgement'. The reference is to Pyrrho of Elis (*c.* 360–270 BC) and his followers.

30 Cf. Matthew 18:3.

reflection follow. And the demonstrations which we employ in the universal process which leads from the senses and things to axioms and conclusions fail us and are incompetent. This process has four aspects, and four faults. First, the impressions of the senses themselves are faulty; for the senses fail and deceive. There need to be substitutions for the failures and corrections for the errors. Secondly, notions are poorly abstracted from sense impressions, and are indeterminate and confused where they should be determinate and sharply defined. Thirdly, induction is poor if it reaches the principles of the sciences by simple enumeration without making use of exclusions and dissolutions, or proper analyses of nature. Finally, the method of discovery and proof, which first sets up the most general principles, and then compares and tests the intermediate axioms by the general principles, is the mother of errors and the annihilation of all the sciences. We are now just touching on these things in passing; we will discuss them at greater length when we explain the true way of interpreting nature, after we have finished these cleansings and purgings of the mind.

LXX

But the best demonstration by far is experience, provided it stays close to the experience itself. For it is a fallacy to pass on to other things which are supposed to be similar, unless the inference is made duly and in order. But the method of experiment[31] which men now use is blind and stupid. Consequently, as they stray and wander in no clear path, just taking their lead from the things they come across, they go round and about and make little headway; sometimes elated and sometimes distracted, they are always finding something else to go on to. Almost always men take their experiences lightly, as if it were a game, making small variations on experiments already known; if the thing does not succeed, they get tired of it, and give up. Even if they take a more serious attitude towards their experiments, more resolute and prepared for hard work, they still devote their efforts to revealing some one experience, as Gilbert did with the magnet and the Chemists with gold. Men act like this because their practice is not merely frivolous but also unintelligent. No search for the nature of a thing is going

---

[31] Or 'experience'. *Experientia* and *experimentum* are used indifferently by Bacon both for the unforced observation which we might call experience and for the contrived experience which we might call an experiment.

to be successful if confined to the thing itself; the inquiry needs to be widened to include more general issues.

If on the basis of their experiments they do construct some kind of science and dogmas, still they nearly always give in to a hasty and premature urge to turn to practical application: not only for the use and profit they may get from such application, but in order to find assurance in the form of a new result that they will not be wasting their time in their future work, and also to advertise themselves to others, in order to increase their reputation in the field of their activities. The result is that, like Atalanta, they go out of their way to pick up the golden apple, and interrupt their running, and let victory slip from their grasp. But in following the true course of experience and directing it towards new results, we should simply take the divine wisdom and order as our example. On the first day of creation God made only the light, and devoted the whole day to this work, and made no material thing that day. We too need first to elicit the discovery of true causes and axioms from every kind of experience: and we must look for illuminating, not profitable, experiments. Once axioms have been rightly discovered and rightly formed, they offer massive assistance to practice. We shall speak later of ways of experiencing which, no less than ways of judgement, have been barred and blocked off; so far we have only said that common experience is a poor demonstration. Now the order of things requires us to say something more about the signs I mentioned earlier, which indicate that the philosophies and observations now in use are inadequate, and on the reasons for a thing which is at first sight so surprising and incredible. Notice of the signs encourages assent; explanation of the reasons removes surprise. These two things greatly assist a swift but gentle purgation of idols from the understanding.

LXXI

Nearly all the sciences we have come from the Greeks. Additions by Roman, Arabic or more recent writers are few and of no great significance; such as they are, they rest on a foundation of Greek discovery. However, the wisdom of the Greeks was rhetorical and prone to disputation, a genus inimical to the search for truth. And so the term 'Sophists', which was rejected by those who wanted to be regarded as philosophers and applied with contempt to the orators - Gorgias, Protagoras, Hippias, Polus – is also applicable to the whole tribe - Plato, Aristotle, Zeno, Epicurus,

Theophrastus and their successors, Chrysippus, Carneades and the rest.[32] The only difference was that the former were itinerant and mercenary, travelling around the cities, making a display of their wisdom and requiring fees; the others were more dignified and more liberal, in that they had fixed abodes, opened schools and taught philosophy without charge. But (though different in other ways) both were rhetorical, and made it a matter of disputations, and set up philosophical sects and schools, and fought for them. Consequently, their teachings were more or less what Dionysius aptly said against Plato – 'the words of idle old men to callow youths'.[33] But the older Greeks[34] – Empedocles, Anaxagoras, Leucippus, Democritus, Parmenides, Heraclitus, Xenophanes, Philolaus and the others (but not the superstitious Pythagoras) – did not, so far as we know, open schools, but gave themselves to the search for truth more quietly, more seriously and more simply, that is with less affectation and display. And therefore, as we suppose, they succeeded better, except that their works have been overwhelmed in the passage of time by those lighter works which are better suited to please the capacity and taste of the crowd; time (like a river) has brought down to us the lighter, more inflated works and sunk the solid and weightier. And yet not even they were completely exempt from the typical vice of their people: they too were too susceptible to the ambition and vanity of founding a sect and winning popular favour. There is no hope for the search for truth when it is sidetracked into these trivialities. And we should not forget, I think, the judgement, or rather prophecy, of an Egyptian priest about the Greeks: 'that they were always children, and had no antiquity of knowledge, nor knowledge of antiquity'.[35] They certainly do have a characteristic of the child: the readiness to talk, with the inability to produce anything; for their wisdom seems wordy and barren of works. And therefore the signs that we gather from the birthplace and family of the philosophy now in use are not good.

---

[32] Gorgias, Protagoras, Hippias and Polus are sophists of the fifth century BC. The rest are a roll-call of the great names of Greek philosophy. Zeno founded the Stoic school in the late fourth century, and Chrysippus, the third head of the school, is its 'second founder'. Epicurus founded the Epicurean school in the late fourth century. Theophrastus succeeded Aristotle as head of the Lyceum. Carneades converted Plato's Academy to Scepticism when he became its head in the mid second century BC.

[33] Dionysius I, tyrant of Syracuse (c. 430–367 BC), as reported at Diogenes Laertius, *Lives of Eminent Philosophers*, III.18.

[34] The pre-Socratics.

[35] Cf. Plato, *Timaeus*, 22B.

LXXII

Nor are the signs which can be gathered from the nature of the time and of the age much better than those from the nature of the place and of the people. Throughout that period knowledge of time and of the world was narrow and limited; and that is a very bad thing indeed, especially for those who stake everything on experience. For they did not have a thousand-year history that deserved the name of history, but fables and rumours of antiquity. They knew only a fraction of the parts and regions of the world, since they called all northern peoples Scythians and all western peoples Celts indiscriminately, knew nothing in Africa beyond the nearest part of Ethiopia, nothing of Asia beyond the Ganges, much less the territories of the New World, even by report or consistent and believable rumour. In fact most climates and zones, in which uncounted nations live and breathe, were declared uninhabitable; and the travels of Democritus, Plato and Pythagoras, which certainly did not take them far from home, were celebrated as major undertakings. But in our time large parts of the New World and the farthest parts of the Old are becoming known everywhere, and the store of experiences has grown immeasurably. Hence if (like astrologers) we are to gather signs from the time of nativity or conception, nothing significant seems to be forecast for those philosophies.

LXXIII

None of the signs is more certain or more worth noticing than that from products. For the discovery of products and results is like a warranty or guarantee of the truth of a philosophy. From these Greek philosophies and the specialised sciences derived from them, hardly a single experience can be cited after the passage of so many years which tends to ease and improve the human condition, and which can be truly credited to the doctrines of the philosophy. Celsus admits it frankly and sensibly: first the experiences of medicine were discovered, and *then* men philosophised about them and sought and assigned causes; it did not happen the other way round, that the experiences were discovered or suggested by philosophy and a knowledge of causes. And so it was no wonder that there were more images of animals than of men among the Egyptians (the Egyptians gave divinity and deification to inventors of things); the reason was that animals have made many discoveries by natural instinct, whereas men have little or none to show as the result of disputation and rational deduction.

But the hard work of the chemists had some results, though more or less by chance and incidentally, or by some variation of the experiments (such as mechanics make), not on the basis of an art or theory; for the fictions they have produced confuse rather than assist experiments. There are few discoveries from those who have worked in so-called natural magic, and they are slight and more like imposture. In religion we are taught that faith is shown by works; and the same principle is well applied to a philosophy, that it be judged by its fruits and, if sterile, held useless; the more so if instead of the fruits of the vine and the olive, it produces the thistles and thorns of disputes and controversy.

LXXIV

Signs should also be gathered from the growth and progress of philosophies and sciences. Those that are founded in nature grow and increase; those founded in opinion change but do not grow. Hence if those doctrines were not completely uprooted like a plant, but were connected to the womb of nature and nourished by her, what we see has been happening now for two thousand years would not have happened: the sciences stand still in their own footsteps and remain in practically the same state; they have made no notable progress; in fact they reached their peak in their earliest author, and have been on the decline ever since. We see the opposite evolution in the mechanical arts, which are founded in nature and the light of experience; as long as they are in fashion, they constantly quicken and grow as if filled with spirit; at first crude, then adequate, later refined, and always progressing.

LXXV

There is another sign also that we should notice (if the term 'sign' properly belongs to it, since it is rather testimony, indeed the most convincing testimony of all); it is the actual confession of the authors whom men now follow. For even those who pronounce on things with such confidence still from time to time, when they come down to earth, resort to complaints about the subtlety of nature, the obscurity of things and the weakness of human understanding. If this was all they did, it might perhaps deter some timid minds from further inquiry, and stimulate and incite men of a sharper and more confident turn of mind to further progress. But it is not enough for them to confess for themselves, but they regard everything that is unknown or untouched by themselves or their masters as beyond the bounds of the possible, and declare, on the authority of their art, that it is impossible for it to be known or done; and so, with supreme arrogance, they turn the weakness of their own discoveries into an insult against nature itself and a vote of non-confidence in other men. Hence the school of the New Academy, which maintained a profession of *lack of conviction* (*acatalepsia*),[36] and condemned men to eternal darkness. Hence the opinion that Forms or the true differences of things (which are in truth the laws of pure act) are impossible to discover and are beyond man. Hence those opinions in the active and operative part of science: that the heat of the sun and the heat of fire are totally different; the fear is of course that men would imagine that through the operations of fire they could extract and form something similar to things that exist in nature. Hence the view that mere composition is a work of man, but mixture is the work of nature alone: in case men should expect from art the generation or transformation of natural bodies. And therefore men will easily allow themselves to be persuaded by this sign not to involve their labours and their fortunes in tenets which are not merely desperate but doomed to despair.

LXXVI

Here is another sign that we must not forget, that there has been so much disagreement among philosophers and such a variety of schools. This reveals clearly enough that the road from the senses to the intellect has not been well built, since the same material of philosophy (namely the nature of things) has been taken and used to construct so many paths of error. And though in our time disagreements and differences of opinion about actual principles and systematic philosophies have been all but extinguished, there still remain countless questions and controversies about the parts of philosophy; thus it is obvious that there is nothing certain or sound in the philosophies themselves or in the modes of demonstration.

[36] See I.37n.

LXXVII

A view exists that there is at least a great consensus around the philosophy of Aristotle, since after its publication the philosophies of the older philosophers allegedly fell into disuse and were forgotten, and in the later period nothing better was discovered; so that it seems to have been so well founded and so well established that it monopolised both periods. But first, the common view of the neglect of ancient philosophy after the publication of the works of Aristotle is not true; long afterwards right down to the time of Cicero and the following centuries the works of the older philosophers survived. It was only in subsequent centuries, when human learning suffered shipwreck, so to speak, from the flood of barbarians that poured into the Roman empire, that the philosophies of Aristotle and Plato were saved by floating on the waves of the times like planks of rather lighter, less solid material. The matter of consensus also is deceptive if one takes a closer look at it. A true consensus is one which (after examination of the matter) consists in liberty of judgement converging on the same point. But the great majority of those who have accepted the philosophy of Aristotle have enslaved themselves to it from prejudice and the authority of others; so that it is rather discipleship and party unity than a consensus. Even if it had been a true, widespread consensus, so mistaken is it that a consensus should be taken for true and sound authority that it implies a strong presumption to the contrary. Worst of all is to take an indication from consensus in intellectual matters, except in divine matters and political matters, where there is a right to vote. For nothing pleases a large number of people unless it strikes the imagination, or confines the mind in the coils of common notions, as was said above. So it is very appropriate to apply Phocion's remark about morals to intellectual matters: men should immediately ask themselves seriously what errors or mistakes they have made if the crowd agrees and applauds.[37] This sign therefore is among the most dangerous. And now we have completed our explanation that the signs of truth and soundness in the philosophies and sciences, as they are now, are poor, whether we gather them from their origins, from their products, from their progress, from their authors' own admissions or from consensus.

---

[37] Phocion, Athenian politician of the fourth century: the saying is reported by Plutarch, *Parallel Lives*, 'Life of Phocion', 8.

LXXVIII

And now we must come to the causes of errors, and the reasons why men have so persistently stuck to them through so many centuries. These causes are many and immensely powerful, so that no one should be surprised that our new approach should have escaped men's notice and remained hidden until now, but rather that they could now at last occur to a man or enter someone's mind. Which is the result more of a kind of good fortune than of any intellectual excellence, and should rather be regarded as a birth of time than an offspring of intelligence.

For, first, if you look at the matter properly, this large number of centuries is reduced to some very short stretches. Of twenty-five centuries in which human memory and learning is more or less in evidence, scarcely six can be picked out and isolated as fertile in sciences or favourable to their progress. There are deserts and wastes of time no less than of regions. We can only count three periods which were high points of learning: one among the Greeks; a second among the Romans; the last among us, the western nations of Europe; and barely two centuries can properly be given to each of them. The middle times of the world were unsuccessful in producing a large or rich crop of sciences. And there is no reason to mention the Arabs or the scholastics, whose numerous treatises in the intervening years rather wore the sciences down than increased their weight. Thus the first reason for such pathetically small progress in the sciences is rightly and properly ascribed to the small amounts of time that have favoured them.

LXXIX

In second place we find a reason which is certainly of great importance throughout, namely that in the periods when human intelligence and letters flourished to a high or even modest degree, natural philosophy occupied a very small part of their efforts. Yet natural philosophy has to be regarded as the great mother of the sciences. For all arts and sciences, torn from this root, may perhaps be polished and fashioned for practical use; but absolutely do not grow. It is clear that after the Christian faith was accepted and grew strong, most of the outstanding intellects applied themselves to theology; and the greatest prizes were offered for this subject, and aids of every kind were copiously provided. This zeal for theology particularly occupied the third part or period of time among us Western Europeans; all the more so as at about the same time letters began to flourish and religious controversies multiplied. In the previous age, while our second period lasted, the period of the Romans, the dominant concerns and labours of the philosophers were engaged and absorbed in moral philosophy (which took the place of theology for the pagans). Also in those times the greatest minds applied themselves for the most part to politics, because of the greatness of the Roman empire, which needed the services of very large numbers of men. The age in which natural philosophy seemed to flourish most among the Greeks was not a great stretch of time, since in the early period the so-called Seven Sages all (except Thales) devoted themselves to moral philosophy and politics; and in the later period, after Socrates had brought philosophy down from heaven to earth, moral philosophy grew still stronger, and turned men's minds away from natural philosophy.

And the very period of time in which the investigation of nature flourished was spoiled and rendered useless by verbal disputes and rivalry in making new dogmas. Thus through these three periods natural philosophy was neglected or obstructed to a pretty large extent, so it is no wonder if men made little progress in the subject, as they were doing something quite different.

LXXX

The next reason is that natural philosophy has scarcely found, even among those who have practised it, anyone to devote his whole time to it, especially in recent times, unless perhaps one might give the example of a monk in his cell or a nobleman at his place in the country 'burning the midnight oil'. Natural philosophy has in fact been treated as a kind of passage or bridge to other things.

That great mother of the sciences with wonderful indignity has been pressed to perform the services of a maid, to minister to the needs of medicine or mathematics, or to wash the immature minds of the young and impregnate them with a kind of first dye, so that later they may absorb some other tincture with greater ease and success. In any case no one should expect great progress in the sciences (especially in their practical part) unless natural philosophy is extended to the particular sciences, and the particular sciences in turn brought back to natural philosophy. This is why astronomy, optics, music, most of the mechanical arts and medicine itself, also (and perhaps more surprisingly) moral and civil philosophy and the sciences of logic, do not reach down to the bottom of things, but simply glide over the variety of things on the surface. For after they have been divided and constituted as particular sciences, they are no longer fed by natural philosophy, which could have lent them new strength and increase from the source, and from true observations of motions, rays and sounds, of the texture and structure of bodies, and of the passions and intellectual processes. It is no wonder that the sciences do not grow when they are cut off from their roots.

LXXXI

And now another important and powerful reason why the sciences have made little progress reveals itself. It is this: it is not possible to get around a racecourse properly if the finishing line is not properly set and fixed. The true and legitimate goal of the sciences is to endow human life with new discoveries and resources. The overwhelming majority of ordinary people have no notion of this, and are concerned only with wages and professional matters; perhaps, occasionally, some unusually intelligent craftsman, seeking to achieve a reputation, devotes himself to making some new invention, usually at his own expense. But in most cases men are so far from setting themselves to augment the sum of sciences and skills that they neither take nor seek more from the sum available than what they can turn to professional use, profit, reputation or similar advantage. And if from so many there is someone who pursues knowledge with genuine love and for its own sake, even he will be found to be pursuing a whole variety of thoughts and doctrines rather than a strict, undeviating investigation of truth. And if there is another perhaps who is a stricter investigator of truth, yet he too will set before himself an account of truth which will satisfy the mind and understanding in providing causes for things already known, not one which will bring results and the new light of axioms. And thus if no one has yet properly set out the end of the sciences, it is no wonder that the consequence in matters subordinate to the end is constant error.

LXXXII

Now just as the end and goal of the sciences is poorly defined among men, so also even if it had been well defined, yet the road which men have chosen for themselves is totally erroneous and impassable. It would strike the mind dumb with amazement, if one thought about it properly, that it has been no one's care or concern that a regular, well-built road should be opened and constructed for the human understanding from sense and experience; but everything has been left to the darkness of traditions, or to the eddy and whirl[38] of arguments, or to the waves and windings of chance and casual, unregulated experience. Let anyone think soberly and carefully about the kind of path men have commonly used in the investigation and discovery of anything. And first he will surely note the simple, non-scientific method of discovery which is most familiar to men. This is simply that in preparing and equipping himself to find something out, anyone first researches and reads what others have written on the subject; then adds his own thoughts, and with much mental agitation interrogates his own spirit and calls upon it to open its oracles to him. This procedure has absolutely no foundation and simply spins around on opinions.

Another person, to aid discovery, might call logic in, which belongs to the subject under discussion only as far as its name is concerned. For a discovery of logic is not a discovery of the leading principles and axioms in which the arts consist, but only of those which seem to be in agreement with them. For the more inquisitive and insistent questioners, those who take the trouble to accost her with demands for proofs and discoveries of principles or of the first axioms, are met by logic with a very well-known response which throws them back on faith and the oath of allegiance (so to speak) which anyone must give to any art.

There remains mere experience: which is chance, if it comes by itself; experiment, if sought. This kind of experience is like a brush without a head (as they say), mere groping, such as men use in the dark, trying everything in case they may be lucky enough to stumble into the right path. It would be much better and more sensible to wait for day or light a lamp, and then to start the journey. The true order of experience, on the other hand, first lights the lamp, then shows the way by its light, beginning with experience digested and ordered, not backwards or random, and from that it infers axioms, and then new experiments on the basis of the axioms so formed; since even the divine word did not operate on the mass of things without order.

Therefore let men cease to wonder that the sciences have not finished

---

[38] 'eddy and whirl' (Kitchin)

the course, since they have wholly lost their way; they have altogether deserted and abandoned experience, or trapped themselves in it (as in a maze) and gone round in circles; since a properly organised order takes one through the woods of experience by a steady path to the open country of axioms.

LXXXIII

The problem has been wonderfully compounded by a certain opinion or judgement, which is deeply ingrained but arrogant and harmful, namely that the majesty of the human mind is diminished if it is too long and too deeply involved with experiments, and with particular things which are subject to the senses and bounded by matter: especially as such things tend to be laborious to investigate, ignoble to think about, crude to speak of, illiberal in practice, infinite in number and short on subtlety. And so it has come at last to this, that the true road is not only deserted but also barred and closed off; experience is badly managed, or despised and even abandoned.

LXXXIV

Again, men have been hindered from making progress in the sciences by the spell (I may say) of reverence for antiquity, and by the authority of men who have a great reputation in philosophy and by the consensus which derives from them. I have spoken above of consensus.[39]

But for antiquity, the opinion which men cherish of it is quite careless, and barely suits the meaning of the word. For true antiquity should mean the oldness and great age of the world, which should be attributed to our times, not to a younger period of the world such as the time of the ancients. True, that age is ancient and older in relation to us, but with respect to the world itself, it was new and younger. We expect from an old man greater knowledge of things human and a more mature judgement than from a young man, both because of his experience and because of the store and variety of things which he has seen and heard and thought about. And truly in the same way it is reasonable that greater things be expected from our age than from old times (if it only knew its strength and was willing to try

[39] See I.77.

68

it and exert it); seeing that our age is the older age of the world, enriched and stocked with countless experiences and observations.

We should also take into account that many things in nature have come to light and been discovered as a result of long voyages and travels (which have been more frequent in our time), and they are capable of shedding new light in philosophy. Indeed it would be a disgrace to mankind if wide areas of the physical globe, of land, sea and stars, have been opened up and explored in our time while the boundaries of the intellectual globe were confined to the discoveries and narrow limits of the ancients.

With regard to authors, it is a mark of supreme cowardice to give unlimited credit to authors and to deny its rights to Time, the author of authors and thus of all authority. For truth is rightly called the daughter of time and not of authority.[40] Therefore it is no wonder if the spell of antiquity, of authors and of consent has so shackled men's courage that (as if bewitched) they have been unable to get close to things themselves.

LXXXV

It is not only admiration for antiquity, authority and consensus which has compelled men's efforts to remain content with things which have already been discovered; but also admiration for the abundance of actual works which have already been provided for the human race. When a man sets before his eyes the variety of objects and the splendid equipment which the mechanical arts have assembled and contributed to human civilisation, he will certainly be more disposed to admire man's wealth than to feel for his poverty; not noticing that men's earliest observations and the operations of nature (which are like the soul and first stirrings of all that variety) are neither many nor profound, and that the rest is due simply to patience and the subtle, ordered movement of hand and tool. The construction of clocks (for example) is certainly a subtle and precise thing that seems to imitate the celestial bodies in its wheels, and the heartbeat of animals in its constant, ordered motion; and yet it depends upon just one or two axioms of nature.

---

[40] Bacon presumably saw the inscription *Veritas temporis filia* (Truth the daughter of time) on the groats of Queen Mary (1553–8). The phrase is also available at Aulus Gellius, *Noctes Atticae*, XII.11. (References from Thomas Fowler (ed.), *Bacon's Novum Organum*, edited with introduction, notes etc. by Thomas Fowler (Oxford, Clarendon Press, 1878). The present translation was made from this edition, which is a critical edition of Bacon's original text published in 1620.)

Again, one might contemplate the subtlety which belongs to the liberal arts, or that involved in the preparation of natural bodies by the mechanical arts, and wonder at such things as the discovery of celestial motions in astronomy, of harmonies in music, of the letters of the alphabet in grammar (not yet in use even now in the kingdom of China); or, again in mechanics, the discovery of the products of Bacchus and Ceres, that is, the making of wine and beer, loaves of bread, or the delicacies of the table and distillation, and so on. One might reflect and notice how many centuries it took for these things to be brought to the state of development which we now enjoy, since all of them, except distillation, were ancient, and (as we said of clocks) how little they owe to observations and axioms of nature, and how easily, by ready opportunities and casual observations, they have been discovered. And thus, I say, he will easily free himself from all wonder, and rather pity the human condition, that through so many centuries there has been such a lack, such a dearth of objects and discoveries. And these discoveries which we have just mentioned are older than philosophy and the arts of under-standing, so that (if the truth be told) since such rational and dogmatic sciences came into being, the discovery of useful works has ceased.

Anyone who has turned his attention from workshops to libraries and conceived an admiration for the immense variety of books we see around us will surely experience a stupendous change of mind once he has given the matter and content of the books themselves a careful examination and inspection. Having observed that there is no end to repetitions, and how men keep on doing and saying the same things, he will pass from admira-tion of variety to amazement at the poverty and paucity of the things which until now have held and occupied the minds of men.

Anyone who lowers himself to contemplate those matters which are thought to be more curious than sensible, and looks deeply into the works of alchemists and magicians, will perhaps be unsure whether they more deserve laughter or tears. The alchemist cherishes an eternal hope, and when a thing does not succeed, he holds his own mistakes responsible. He accuses himself of not properly understanding the words of the art or of the authors, and so proceeds to give his attention to traditions and underground reports; or of making a slip in the weights or timing of his procedure, and so proceeds to repeat the experiment indefinitely; and in the meantime when among the hazards of his experiments he chances upon something new in appearance or usefully worth saving; he feeds his spirit on such promises, and exaggerates and advertises them: for the rest he lives in hope.

And yet it must not be denied that alchemists have discovered quite a few things, and given men useful discoveries. They fit quite well in the story of the old man who left his daughters some gold buried in a vineyard and pretended not to know the exact spot; as a result of which they set themselves to dig diligently in that vineyard; and no gold was found, but the harvest was richer for the cultivation.

But the devotees of natural magic, who deal with everything by the sympathies and antipathies of things, have ascribed wonderful powers and effects to things on the basis of idle, lazy conjectures; and any results they have shown are good only for surprise and novelty, not profit and usefulness.

In superstitious magic (if we must speak of this too) we should particularly notice that among all nations and religions in all times, it is only in subjects of a certain limited kind that the curious and superstitious arts have done, or seem to have done, anything; so let us dismiss them. To sum up, it is not surprising that our belief that we have a lot has been a cause of our having so little.

LXXXVI

Men's wonder at learning and the arts, which is simple enough in itself and almost like the wonder of children, has been reinforced by the cunning and artifice of those who have practised and taught the sciences. They bring them forward with much show and affectation, and put them before the public in a misleading and disguised fashion, in order to give the impression that they are quite finished and honed to the last degree of perfection. For if you look at their method and their divisions, they appear to embrace and include absolutely everything that can come under that subject. True, the subdivisions are empty, like bookshelves without books, but to the common mind they have the form and organisation of a complete science.

But the earliest and most ancient investigators of truth, with better credit and success,[41] used to cast the knowledge which they set themselves to gather from the contemplation of things and to store for use, into the form of *aphorisms*, or short, unconnected sentences, not methodically arranged; and did not pretend or profess to possess a universal art. But as things are done now, it is no wonder at all if men do not look any further into subjects that are taught as complete and long perfected in all their parts.

[41] The Greek pre-Socratic philosophers. Cf. 1.63 and note.

LXXXVII

The old ideas have also received a big boost to their reputation and credit from the slick and empty claims made by supporters of the new, especially in the practical and applied part of natural philosophy. There have been plenty of glib, fanciful talkers who have showered the human race with promises, partly from credulity and partly from imposture: promising and advertising longer life, postponement of old age, relief from pain, healing of natural defects, temptations for the senses, enchantment and excitement of the passions, stimulation and enlightenment of the intellectual faculties, transmutation of substances, unlimited power and variety of movement, impressions and alterations of air, drawing and control of celestial influences, divination of future things, representation of distant things, revelation of hidden things and much more of the same. The right verdict on these false benefactors is that in philosophical teaching there is just as much difference between their empty promises and the true arts as there is, in the narratives of history, between the achievements of Julius Caesar or Alexander the Great and the deeds of Amadis of Gaul or Arthur of Britain.[42] Those famous generals are found to have actually done greater things than those fictitious heroes are even feigned to have done; and in ways and manners of action which are not at all fabulous and marvellous. Nor is it reasonable that the credit of true memory should be diminished because it has sometimes been damaged and violated by fables. But in any case it is no wonder that the impostors who have tried such things have raised a great prejudice against new proposals (especially when they are said to have practical effects), since their excessive vanity and contempt even in our day has destroyed all belief in efforts of that kind.

LXXXVIII

Much greater harm has been done to the sciences by lack of ambition and by the pettiness and poverty of the projects that human industry has set itself. And, worst of all, this lack of ambition is accompanied by arrogance and contempt.

---

[42] Julius Caesar (?102–44 BC), Roman politician, general and dictator, conqueror of Gaul. Alexander of Macedon (356–323 BC), conqueror of the Persian empire. Amadis of Gaul, fictional hero of the medieval romance of that name, which was popular down to Bacon's own time. Arthur of Britain: legendary hero of the Arthurian cycle of stories.

First there is the excuse that has become familiar in all the arts whereby authors turn the feebleness of their art into a false accusation against nature; and what their art fails to achieve, they declare, on the basis of the same art, to be impossible in nature. And certainly an art cannot be condemned if it is its own judge. Even the philosophy which is now current cherishes some positions or doctrines in its bosom which (if you look carefully) they want everyone to accept without question: that nothing should be expected from art or the work of man which is difficult or masterful or strong against nature; as we said above about the difference in kind between the heat of a star and the heat of fire, and about mixture. If you look at it carefully, this is wholly due to a wilful limiting of human power, and to an artificially manufactured desperation, which not only dims any visions of hope, but also blights all the incentives and nerves of industry, and rejects the hazards of experience itself. They only care that their art should be thought to be perfect; and they make the vainest and most damaging pretense that whatever has not yet been discovered and understood cannot be expected to be found or understood in the future. If however someone does try to devote himself to things and to find something new, he will simply decide to set himself to make a thorough and detailed investigation of some one discovery (and no more): the nature of the magnet, for example, the ebb and flow of the tides, the system of the sky, and so on, which seem to hold some secret and so far have been treated with little success. It is a mark of the highest incompetence to make a thorough investigation of the nature of some one thing by itself, seeing that the same nature which seems to be latent and hidden in some things is manifest and graspable in others, and is a cause of wonder in the first case but in the latter does not even draw attention. This is the case with the nature of consistency, which in wood or rock is not particularly noted but casually referred to as solid, without further inquiry about their refusal of separation or dissolution of their continuity; but in the case of bubbles of water the same thing seems to be subtle and ingenious, because they wrap themselves in little films, curiously shaped in the form of a hemisphere, so that for a moment of time they avoid dissolution of their continuity.

Moreover, even things which are thought to be secret have an open and public nature in other cases; which will never allow itself to be seen, if men's experiments or thoughts are only concerned with the things themselves. And it is generally and commonly taken to be a new discovery in mechanical effects if one puts a subtler finish on things discovered long before or dresses them out more elegantly, or puts things together and unifies them, or ties them more conveniently to their use, or presents an effect with a greater or even a smaller amount or volume than it was accustomed to have, and so on.

Therefore it is no wonder that no remarkable discoveries worthy of the human race have been brought into the light, since men have been happy and content with such little, childish tasks; and in fact have supposed that in them they were pursuing or achieving something important.

LXXXIX

Nor should we fail to mention that in every age natural philosophy has had a troublesome and difficult adversary, namely superstition and the blind, immoderate zeal of religion. One may see among the Greeks that those who first proposed natural explanations for lightning and storms to men who had never heard of such a thing were found guilty of impiety against the gods. Not much better treatment was given by some of the ancient fathers of the Christian religion to those who on the basis of most evident proofs (which today no sane man would contradict) took the position that the earth is round, and in consequence asserted that the antipodes exist.

Furthermore, as things are now, the situation for discussions of nature has been made much harder and more dangerous by the *Summaries* and *Methodical Treatises* of the scholastic theologians, who when they reduced theology to order (as much as they could) and fashioned it in the form of an art, also succeeded in mixing the prickly and contentious philosophy of Aristotle more thoroughly into the body of religion than was appropriate.

The same tendency is shown (though in a different manner) by the treatises of those who have not been afraid to deduce and confirm the truth of the Christian religion from the principles and authority of philosophers. With much pomp and ceremony they celebrate the marriage of faith and sense as a legitimate union, and charm men's minds with a pleasing variety of things, but at the same time mix things human with things divine, an unequal union. Such mixtures of theology and philosophy find room only for what is currently acceptable in philosophy; new things, though a change for the better, are all but dismissed and excluded.

Finally you will find that some theologians in their ignorance completely block access to any philosophy, however much emended. Some are simply anxious that a closer investigation of nature may penetrate beyond the

permitted boundaries of sound opinion; they misinterpret what the holy
Scriptures, in talking of divine mysteries, have to say against prying into
God's secrets, and wrongly apply it to the hidden things of nature, which
are not forbidden by any prohibition. Others, more cunningly, conjecture
and imagine that if the intermediate causes of things are unknown,
individual events can more easily be attributed to the hand and rod of God
(which is, as they suppose, very much in the interest of religion); this is
simply an attempt 'to please God by a lie'.[43] Others fear from example that
movements and changes in philosophy will invade religion and settle
there. Others, finally, seem anxious that something might be found in the
investigation of nature which would undermine or at least weaken religion
(especially among the uneducated). These two latter fears seem to us to
smell too much of carnal wisdom, as if men had no confidence and felt
doubt in the depths of their mind and in their secret thoughts about the
strength of religion and the dominion of faith over the senses; and were
therefore afraid and felt threatened by the investigation of truth in natural
matters. But truly, if one thinks about it, natural philosophy, after the word
of God, is the strongest remedy for superstition and the most proven food
of faith. Therefore it has deservedly been granted to religion as its most
faithful handmaid; for one manifests the will of God, the other his power.
He was not mistaken who said: 'ye do err, not knowing the scriptures and
the power of God',[44] mixing and uniting the revelation of his will and the
thought of his power in an indivisible bond. No wonder the growth of
natural philosophy has been inhibited, since religion, which has the most
enormous power over men's minds, has been kidnapped by the ignorance
and reckless zeal of certain persons, and made to join the side of the enemy.

XC

Again, in the manners and customs of the schools, universities, colleges and
similar institutions, which are intended to house scholars and cultivate
learning, everything is found to be inimical to the progress of the sciences.
For the readings and exercises are so designed that it would hardly occur
to anyone to think or consider anything out of the ordinary. And if perhaps
someone should have the courage to use his liberty of judgement, he would
be taking the task on himself alone; he will get no useful help from his

[43] Cf. Job 13:7.
[44] Matthew 22:29.

colleagues. And if he puts up with this too, he will find that in pursuing his career his industry and largeness of view will be no small obstacle to him. For men's studies in such places are confined and imprisoned in the writings of certain authors; anyone who disagrees with them is instantly attacked as a troublemaker and revolutionary. But there is certainly a great difference between political matters and the arts: there is not the same danger from a new movement and a new light. In political matters even a change for the better is suspected as subversive, since politics rests upon authority, consent, reputation and opinion, not on demonstration. But in the arts and sciences, as in a mine, everything should hum with new works and further progress. This is how things are according to right reason, but it is not how things actually are; the administration and policy towards learning which we have discussed has long had a gravely suppressive effect on the advancement of the sciences.

XCI

Furthermore, even if this jealous opposition stopped, it is still enough to check the growth of the sciences that such efforts and labours are without reward. For the cultivation of the sciences and reward are not in the same hands. For the growth of the sciences comes inevitably from great intellects; but the prizes and rewards of the sciences are in the hands of the vulgar or of princes, who (with very rare exceptions) are barely even moderately learned. In fact such advances fail to obtain not only rewards and goodwill from men but even popular praise. For they are beyond the understanding of the majority of men, and are easily overwhelmed and extinguished by the winds of popular opinion. No wonder that a thing which has not been honoured has not prospered.

XCII

But much the greatest obstacle to the progress of the sciences and to opening up new tasks and provinces within them lies in men's lack of hope and in the assumption that it is impossible. For grave and prudent men tend to be quite without confidence in such things, reflecting in themselves on the obscurity of nature, the shortness of life, the defects of the senses, the weakness of judgement, the difficulties of experiments, and so on. And so they suppose that there is a kind of ebb and flow of knowledge, through the turnings of time and the ages of the world; since in some periods they grow and flourish, in others decline and fall: and always under this law, that when they have reached a certain level and condition, they can go no further.

Therefore, if anyone believes or promises more, they think it is a sign of an immoderate and immature mind; and believe that in such efforts the beginning is happy, the middle is difficult and the end confusion. And since such thoughts are those which readily occur to serious men of superior judgement, we must truly be careful not to be captivated by our love for the best and most beautiful thing, and relax or lessen the severity of our judgement. We must carefully consider what hope appears and from what direction it comes; we must reject the lighter breezes of hope, and thoroughly analyse and weigh those which seem most sound. We also need to call in political prudence to give advice, and we must make use of it; it is distrustful by its nature, and takes a dim view of human affairs. And so we must now speak of hope; especially as we do not make rash promises nor do violence to men's judgments, nor lay snares for them, but lead them by the hand and with their own consent. By far the most powerful remedy for impressing hope would be to lead men to particulars, especially as digested and arranged in our tables of discovery (these belong partly to the Second but mostly to the Fourth Part of our *Renewal*), since this is not mere hope but the thing itself. Yet to do all gently, we must proceed in our plan of preparing men's minds; and the display of hope is no small part of that preparation. For without hope, the rest of it tends to make men sad (i.e. it gives men a worse and lower opinion than they had of things as they are now, and sharpens their perceptions and feelings about the poverty of their condition) rather than lend them any eagerness or sharpen their keenness to experiment. And therefore we should reveal and publish our conjectures, which make it reasonable to have hope: just as Columbus did, before his wonderful voyage across the Atlantic Sea, when he gave reasons why he was confident that new lands and continents, beyond those previously known, could be found; reasons which were at first rejected but were afterwards proven by experience, and have been the causes and beginnings of great things.

XCIII

We must begin from God, since our work, because of the supreme element of good in it, is manifestly from God, who is the author of good and the father of lights.[45] And in the operations of God, the beginnings, however slender, have a sure end. What is said of spiritual things, 'the kingdom of God does not come with watching'[46] is also found to be true in all the greater works of providence; so that all things move smoothly without sound or commotion, and the thing is wholly done before men realise or notice that it is under way. Nor should we omit the prophecy of Daniel on the last times of the world: 'Many shall pass through, and knowledge will be increased',[47] which obviously signifies enigmatically that it is in the fates, that is, in providence, that the circumnavigation of the world (which after so many long voyages now seems quite complete or on the way to completion) and the increase of the sciences should come to pass in the same age.

XCIV

Now comes the reason which most of all gives grounds for hope: namely from the errors of past times and of the ways so far tried. There is a very fine reproach about the poor handling of a political situation which some-one expressed in these words: 'What is worst in the past should seem the best for the future. For if you had done everything which belongs to your duty, and your affairs were in no better shape, there would be no hope left that they could take a turn for the better. But since your affairs are bad, not through force of the situation but by your own mistakes, it is to be hoped that when those mistakes are put aside or corrected, a great change for the better may be made in your affairs'.[48] In a similar way, if through so many periods of years men had kept to the true way of discovering and develop-ing the sciences, and had not been able to progress further than they have, it would certainly be a bold and rash opinion that the thing could advance any further. But if they had mistaken the way, and men's efforts were spent on things on which they should not have been spent, it follows that the source of the difficulty is not in the things themselves, which are not in our power, but in the human understanding, and its use and application; and that is susceptible of remedy and cure. Therefore the best thing would be to set out these errors: for every error that has been an obstacle in the past

[45] Cf. James 1:17.
[46] Luke 17:20.
[47] Daniel 12:4. This is the motto on the title page of the 1620 edition of *The New Organon*.
[48] Demosthenes, *Philippic*, III.4; cf. *Philippic*, I.2. Athenian orator of the fourth century BC.

is an argument of hope for the future. Though these obstacles have been lightly touched on in what we said above, still it has also seemed good to present them briefly now in plain and simple words.

XCV

Those who have treated of the sciences have been either empiricists or dogmatists. Empiricists, like ants, simply accumulate and use; Rationalists, like spiders, spin webs from themselves; the way of the bee is in between: it takes material from the flowers of the garden and the field; but it has the ability to convert and digest them. This is not unlike the true working of philosophy; which does not rely solely or mainly on mental power, and does not store the material provided by natural history and mechanical experiments in its memory untouched but altered and adapted in the intellect. Therefore much is to be hoped from a closer and more binding alliance (which has never yet been made) between these faculties (i.e. the experimental and the rational).

XCVI

Natural philosophy is not yet found in a pure state, but contaminated and corrupted: in the school of Aristotle by logic, in the school of Plato by natural theology; in Plato's second school, that of Proclus and others, by mathematics, which should only give limits to natural philosophy, not generate or beget it. Better things are to be hoped from natural philosophy pure and unadulterated.

XCVII

There is no one yet found of such constancy and intellectual rigour that he has deliberately set himself to do completely without common theories and common notions, and apply afresh to particulars a scoured and level intellect. And thus the human reason which we now have is a heap of jumble built up from many beliefs and many stray events as well as from childish notions which we absorbed in our earliest years.

But if someone of mature age, with faculties unimpaired and mind cleansed of prejudice, applies himself afresh to experience and particulars, better is to be hoped of him. And in this task we promise ourselves the fortune of Alexander the Great; and let no one accuse us of vanity before he sees the result of the thing, which aims to uproot all vanity.

For of Alexander and his achievement Aeschines spoke as follows: 'we certainly do not live a mortal life; but we were born for this, that posterity should say and proclaim marvels of us':[49] exactly as if he regarded Alexander's achievements as miraculous.

In later times Titus Livy considered the matter and saw more deeply into it, and spoke somewhat as follows about Alexander:[50] 'that he simply had the courage to despise vanities'. And we think that the same judgement will be made of us in future times: 'that we did nothing great, but simply put less value on things which are held to be great'. But meanwhile (as we have said already) there is no hope except in the *renewal* of the sciences, i.e. that they may be raised up in a sure order from experience and founded anew; which no one (we think) would affirm has yet been done or contemplated.

XCVIII

The foundations of experience (since we absolutely must get down to this) have been non-existent or very weak; nor has a collection or store of particulars yet been sought or made, able or in any way adequate, either in number, kind or certainty, to inform the intellect. To the contrary, learned men (admittedly idle fellows and easy-going) have accepted in the formation or confirmation of their philosophy some reports, or rather rumours and whispers, of experience, and have given them the weight of legitimate testimony. Imagine a kingdom or state basing its counsels and business not on letters and reports from ambassadors and messengers worthy of credence, but on the gossip of men about town and trivialities; this is just the kind of administration that has been brought into philosophy, as far as experience goes. Natural history contains nothing that has been researched in the proper ways, nothing verified, nothing counted, nothing weighed, nothing measured. But what is indefinite and vague in observation is deceiving and unreliable as information. And if this seems to anyone a strange thing to say and an unjustified complaint, since Aristotle, so great a man himself and supported by the resources of so great a king, achieved such an accurate history of animals, and others with greater diligence (but

---

[49] Aeschines, *Against Ctesiphon*, 132. Athenian orator of the fourth century BC, supporter of Alexander of Macedon and political opponent of Demosthenes.

[50] Livy, *History of Rome* [*Ab Urbe condita*], IX.17 *ad fin.*

less noise) have made many additions, and others again have composed copious histories and narratives of plants, of metals and of fossils, he seems not to be paying proper attention and seeing what is at stake here. The method of a natural history which is made for itself is one thing; quite other is the method of natural history which is gathered to inform the understanding in order to found a philosophy. These two histories differ in many ways, but especially in this, that the first of them contains a variety of natural species, not experiments of the mechanical arts. For just as in politics each man's character and the hidden set of his mind and passions is better brought out when he is in a troubled state than at other times, in the same way also the secrets of nature reveal themselves better through harassments applied by the arts than when they go on in their own way. And so we have our best hope of natural philosophy once natural history (which is its base and foundation) has been better organised; but not before.

XCIX

And again the very wealth of mechanical experiments reveals the supreme poverty of the things which most help and assist the information of the understanding. For a mechanic, who is by no means anxious about the investigation of truth, does not direct his mind or stretch his hand to anything but what is useful for his task. But the hope of further progress in the sciences will be well founded only when natural history shall acquire and accumulate many experiments which in themselves are of no use, but which simply help towards the discovery of causes and axioms; these we have been accustomed to call *illuminating* experiments as distinct from *profitable* experiments. And they have in themselves a wonderful power and provision, namely that they never deceive or disappoint. For since they are used not to make a product but to reveal the natural cause in something, they equally answer to their intention, however they turn out; since they put an end to the question.

C

Thus we must seek to acquire a greater stock of experiments, and experiments of a different kind than we have yet done; and we must also introduce a quite different method, order and process of connecting and advancing experience. For casual experience which follows only itself (as we said above) is merely groping in the dark, and rather bemuses men than informs them. But when experience shall proceed by sure rules,[51] serially and continuously, something better may be expected from the sciences.

CI

But even when you have available and ready the stock and material of natural history and experience that is necessary to the work of the intellect, or philosophic work, the intellect is still quite unable to work on the material on its own and by memory; as if one expected to be able to memorise and master the calculations of an account book. Yet up to now the role of thought has been more prominent than that of writing in the work of discovery; no *written experience* has yet been developed, though we should not approve any discovery unless it is in writing. When it shall come into use, we may expect more from experience finally made literate.

CII

Moreover, since there are so vastly many particulars (a whole *host* of particulars), and since they are so scattered and diffuse that they distract and confound the understanding, we should not expect much from its casual and undirected motions and cursory movements unless we introduce arrangement and coordination by appropriate, well-organised and living (so to speak) tables of discovery of things relevant to the subject of the investigation, and apply our minds to the organised summaries of facts which these tables provide.

CIII

But after a store of particulars has been put before us in the due and proper order, we should not go straight on to investigate and discover new particulars or effects, or if we do, we must not stop there. For after all the experiments of all the arts have been collected and digested and have been brought before the notice and judgement of one man, we do not deny that from the actual transfer of experiments from one art to another, many new things may be found which will be useful for human life and its condition,

[51] *lege certa*

by means of the experience which we call *written experience*. And yet only small things may be hoped from it; the more important are to be expected from the new light of axioms drawn by a sure method and rule from particulars, which may in turn indicate and point to new particulars. For the road is not flat, but goes up and down - up first to the axioms, then down to the effects.

CIV

But we must not allow the understanding to leap and fly from particulars to remote and highly general axioms (such as the so-called *principles* of arts and things), and on the basis of *their* unshakable truth, demonstrate and explicate the intermediate axioms, as is still done, since the mind's natural bent is prone to do this, and is even trained to it and made familiar with it by the example of syllogistic demonstration. But one may only expect any-thing from the sciences when the ascent is made on a genuine ladder, by regular steps, without gaps or breaks, from particulars to lesser axioms and then to intermediate axioms, one above the other, and only at the end to the most general. For the lowest axioms are not far from bare experience. And the highest axioms (as now conceived) are conceptual and abstract, and have no solidity. It is the intermediate axioms which are the true, sound, living axioms on which human affairs and human fortunes rest; and also the axioms above them, the most general axioms themselves, are not abstract but are given boundaries by these intermediate axioms.

Therefore we do not need to give men's understanding wings, but rather lead and weights, to check every leap and flight. And this has not been done before; but when it shall be done, we may have better hope of the sciences.

CV

In forming an axiom we need to work out a different form of induction from the one now in use; not only to demonstrate and prove so-called principles, but also lesser and intermediate axioms, in fact all axioms. For the induction which proceeds by simple enumeration is a childish thing, its conclusions are precarious, and it is exposed to the danger of the con-trary instance; it normally bases its judgement on fewer instances than is appropriate, and merely on available instances. But the induction which will be useful for the discovery and proof of sciences and arts should separate out a nature, by appropriate rejections and exclusions; and then, after as many negatives as are required, conclude on the affirmatives. This has not yet been done, nor even certainly tried except only by Plato, who certainly makes use of this form of induction to some extent in settling on definitions and ideas. But any number of things need to be included in a true, legitimate account of this kind of induction or demonstration, which have never occurred to anyone to think about, so that more effort needs to be put into this than has ever been spent on the syllogism. It is this kind of induction whose help we must have not only to discover axioms but also to define concepts. And we may certainly have the greatest hopes for this kind of *induction*.

CVI

In forming axioms by this kind of induction we need also to conduct an examination and trial as to whether the axiom being formed is only fitted and made to the measure of the particulars from which it is drawn, or whether it has a larger or wider scope. If it is larger and wider in scope, we must see whether, like a kind of surety, it gives confirmation of its scope and breadth by pointing to new particulars; so that we do not just stick to things that are known, nor on the other hand extend our reach too far and grasp at abstract forms and shadows, not at solid things clearly defined in the material. When these axioms come into use, then at last a well-founded hope will truly have appeared.

CVII

And here we must also repeat what we said above about the enlargement of natural philosophy and the relation[52] of the particular sciences to it, so that there will be no division and dismemberment of the sciences; without this, little progress may be expected.

CVIII

And now we have finished speaking about the abolition of despair and the acquisition of hope by dismissing or correcting the errors of the past. We

---

[52] 1.79, 80.

must now see if there are any other things that give hope. And this comes to mind: if many useful things have been found by lucky accident or happy opportunity when men were not looking for them but engaged on something else, no one can doubt that when they do look and attend to this thing and not something else, and when they do so with method and order, not impulsively and desultorily, many more things are bound to be uncovered. For though it may happen now and again that someone may fall by pure luck upon something which had evaded him before, when he was making a great effort and putting a lot of work into investigating it, nevertheless in general the opposite is undoubtedly the case. Therefore many more things, better things, and at more frequent intervals, are to be hoped from human reason, hard work, direction and concentration than from chance, animal instinct and so on, which are what up to now have been the origin of discoveries.

CIX

This too might be cited as ground for hope, that some of the things discovered in the past were such as no one would be likely to have any inkling of before they were discovered; anyone would have flatly rejected them as impossible. For men are accustomed to divine the new by the example of the old, and by an imagination schooled and stained by the old; which is the most deceptive kind of thinking, seeing that much that is drawn from the sources of things does not flow through the usual channels.

If before the discovery of the cannon one had described the thing by its effects, and said something like this: 'a discovery has been made, by which the biggest walls and fortifications may be smashed and thrown down from a great distance', men would surely have been likely to have many different ideas about increasing the force of catapults and siege-engines by means of weights and wheels and similar mechanisms for battering and striking. But a fiery wind so suddenly and violently expanding and exploding would have been unlikely to occur to anyone's imagination or fancy; since he would not have seen an example of these at first hand, except perhaps in an earthquake or thunderbolt, which men would immediately have been likely to reject as monstrous forces of nature not imitable by human beings.

In the same manner, if before the discovery of the silkworm's thread someone had remarked: 'a kind of thread has been discovered which can be used for clothes and furnishings, which is much finer than linen or wool thread and yet stronger, as well as softer and more lustrous', men's first thoughts would have been about some silken plant, or the delicate coat of an animal, or bird feathers or down; but the web of a little worm and one which is so productive and renews itself every year - they would surely never come up with that. And if anyone had suggested a worm, he would certainly have been made fun of for dreaming of a new kind of cobweb.

Similarly, if before the invention of the mariner's compass anyone had remarked that an instrument has been invented by which the poles and the points of the sky may be taken and distinguished with precision, men would have got their imaginations to work on it and started talking about the more precise construction of astronomical instruments in many different ways, but it would have seemed altogether incredible that anything would be found whose motion agreed so well with the heavenly bodies, not being itself a heavenly body but only a stone or metal substance. Yet this and similar things through so many ages of the world have been hidden from men, and were not discovered by philosophy or the mechanical arts but by chance and accident; and they are of such a kind (as we said before) that they are completely different and remote from things previously known, so that no prior conception[53] would ever have led to them.

Therefore it is very much to be expected that many exceedingly useful things are still hidden in the bosom of nature which have no kinship or analogy with things already discovered, but lie altogether outside the paths of the imagination; which however have not yet been discovered; but without a doubt will appear sometime, through the many twists and turnings of the centuries, just as the discoveries mentioned above appeared. However, if we follow the way we are now discussing, they may be exhibited and anticipated speedily, suddenly and at once.

CX

Other discoveries too are seen to confirm that the human race can miss and ignore remarkable discoveries even when they are lying at their feet. For though the inventions of gunpowder, silk from the silkworm, the mariner's compass, or sugar or paper or similar things may seem to depend upon certain properties of things and nature, yet the technique of printing

---

[53] *praenotio*: Case refers to Greek *prolepsis* (a Stoic notion) and suggests 'preconception'.

certainly contains nothing which is not open and almost obvious. And yet men went without this magnificent invention (which does so much for the spread of learning) for so many centuries, because they did not notice that though it is obviously more difficult to place letter types than to write letters by movements of the hand, it has the advantage that once letter types are placed they are good for an infinite number of printings, whereas letters set down by hand are good only for the one script; or perhaps because they did not notice that ink can be thickened so that it marks without running, especially if the letters are face up and the impression is made from above.

In this course of discovery the human mind is so often accustomed to be awkward and unskilful that at first it lacks confidence and soon comes to despise itself; and at first it seems incredible to it that such-and-such a thing can be discovered, but after it has been discovered, it again seems incredible that it could elude men for so long. And this itself is rightly given as grounds for hope, namely that a great store of inventions remains to be found which may be brought out by what we call *written experience*, not only by unearthing unknown operations but also by transferring, compounding and applying operations already known,

CXI

Nor should we omit the following as a reason for hope. Think (if you will) of the infinite expenditure of talent, time and resources which men invest in things and pursuits of much less use and value. If even a fraction were turned to solid and sensible subjects, any difficulty could be overcome. The reason why we decided to add this is that we plainly confess that such a collection of experimental and natural history as we see in our mind's eye and as it should be, is a great work, I may say a royal work, and a work of much effort and expense.

CXII

Let no one be frightened of the sheer number of particulars; in fact this very thing should restore his hope. For the particular phenomena of the arts and nature are a mere handful in comparison with the mind's fictions when they are abstracted and lose their connection with the evidence of things. And the end of this route is plain, and almost in the vicinity; the other has no end, it is an unending maze. For men have not yet spent much time on experience, they have only grazed it lightly, but they have wasted an infinite time on cogitation and intellectual gymnastics. If there were anyone present among us who would answer interrogatories about the facts of nature, it would take only a few years to discover all causes and all sciences.

CXIII

We also think that men might take some encouragement from our example; and we do not say this to boast, but because saying it is useful. Anyone who is diffident should see how I have (as I think) advanced the subject to some degree, though I am the busiest man of my age in political affairs, not in very good health (which wastes a great deal of time), and a genuine pioneer in this domain,[54] not following in anyone's footsteps, nor sharing these thoughts with any other human being, and yet constantly walking in the true way and submitting my mind to nature. Then let them see from these tokens we give what may be expected from men who have abundance of leisure, and from cooperative labours and from the passage of time; especially on a road which may be travelled not only by individuals (as is the case in the *way of reason*), but where men's labours and efforts (particularly in the acquisition of experience) may be distributed in the most suitable way and then reunited. For men will begin to know their own strength when we no longer have countless men all doing the same thing, but each man making a different contribution.

CXIV

Lastly, even if the breeze of hope blew much more weakly and faintly from *this New Continent*, still we believe that the attempt has to be made (unless we want to be utterly despicable). For the danger of not trying and the danger of not succeeding are not equal, since the former risks the loss of a great good, the latter of a little human effort. But from what we have said and from other things which we have not said, it has seemed to us that we have abundance of hope, whether we are men who press forward to meet new experiences, or whether we are careful and slow to believe.

---

[54] 'pioneer': Bacon uses *protopirus*, a transliteration of an unusual Greek word which he might possibly have found in Polybius, 1.61.4, through Polybius' editor, Casaubon, whom Bacon invited to England and spent time with.

CXV

And now we have finished speaking of the removal of despair, which has been among the most powerful causes which have delayed and retarded the progress of the sciences. And we have completed our discussion of the signs and causes of error and of the prevailing inertia and ignorance; its subtler causes, beyond the scope of popular judgement or observation, should be related to what has been said about the Idols of the mind.

This is also the end of the destructive part of our Instauration. It has consisted of three Rebuttals: the rebuttal of *Native Human Reason* left to itself; the rebuttal of *Demonstrations*; and the rebuttal of *Theories*, or the received philosophies and commonly accepted doctrines. Our rebuttal of them has been such as it could be, namely by signs and the evidence of causes; this is the only form of refutation available to us (since we disagree with others about principles and forms of proof).

And therefore it is time for us to approach the actual art and norm of *Interpreting Nature*; and yet a preliminary note still needs to be made. For our purpose in this first book of Aphorisms has been to prepare men's minds both to understand and to accept what follows; now that the platform[55] of the mind has been scraped and made level, the next step is to set it in a good position, with a favourable view towards what we shall exhibit. For in a new business prejudice may be caused not only by the influence of a powerful old belief, but also by a false preconception or unwarranted image of the new thing being offered. And therefore we shall try to ensure that good and true opinions are held about the things we are introducing, even if they are only for the time being, like an interest payment, until the thing itself may be seen more clearly.

CXVI

First then it seems we must ask men not to suppose that we are trying to found a new sect in philosophy, in imitation of the style of the ancient Greeks or of certain moderns, such as Telesio, Patrizzi or Severinus.[56] This

---

55 'platform' (Kitchin)

56 Bernardino Telesio (1509-88), author of *De rerum natura juxta propria principia* [On the nature of things according to their proper principles] (1565–86), anti-Aristotelian; Francesco Patrizzi (1529–97), 'the most systematic of the opponents of the Aristotelian philosophy in general' (Fowler, 'Introduction', p. 84), author of *Discussiones peripateticae* [Peripatetic discussions]; 'Severinus' may be either Petrus Severinus (1542–1602) or M. A. Severinus (1580–1656).

is not what we are doing, and we do not believe that it makes much difference to men's fortunes what sort of abstract opinions anyone has about the nature and principles of things. There is no doubt that many such old opinions might be revived and new ones introduced, just as we may entertain several hypotheses about the heavens which are more or less compatible with the phenomena but incompatible with each other.

We are not working on such matters of opinion, useless things. On the contrary, our design is to discover whether in truth we can lay firmer foundations for human power and human greatness, and extend their limits more widely. Here and there, in some special subjects, we have much truer, more certain foundations (we believe) which are also more profitable than those which men now use (and we have collected them in the fifth part of our Renewal); nevertheless we are not proposing any universal or complete theory. This does not seem to be the time for that. Moreover, we do not expect to live long enough to complete the sixth part of our Renewal (which is dedicated to the discovery of philosophy through the legitimate interpretation of nature). We are content if we conduct ourselves soberly and usefully in the middle parts, and scatter throughout the seeds of more complete truth for posterity, and not falter at the commencement of great things.[57]

CXVII

Thus we are not founders of a sect; nor are we benefactors or promisers of particular results. But one might make this request, that we who so often speak of results, and relate everything to that end, would also give some samples of them. But our way and method (as we have often said clearly, and are happy to say again) is not to draw results from results or experiments from experiments (as the empirics do), but (as true Interpreters of Nature) from both results and experiments to draw causes and axioms, and from causes and axioms in turn to draw new results and experiments.

It is true that in our tables of discovery (which is the content of the fourth book of the Instauration), in the examples of particular things which we have given in the second part, and also in our observations on history, described in the third part, even a person of average insight and

---

[57] Bacon did not live to complete any of these projected parts of *The Great Renewal*, though some surviving short pieces were probably intended to be included in the great work.

intelligence will see here and there indications and pointers to several remarkable results; but we openly admit that the natural history which we have at this day, either from books or from personal examination, is not so full and so properly verified that it can satisfy or serve legitimate Interpretation.

Hence we give leave and permission to anyone who is better suited to mechanical things, and better trained, and ingenious in deriving results from mere acquaintance with experiments, to undertake the difficult task of gathering a good crop from our history and from our tables as he passes by, taking an interest payment for the time being until the capital can be had. But we have a larger goal, and condemn all untimely and premature activity of this kind as balls of Atalanta (as we like to call them). We do not grasp at golden apples like a child; but stake the whole race on the victory of art over nature; we are in no hurry to collect moss or cut the green corn; we wait for the harvest to be ready.

CXVIII

It will undoubtedly also occur to someone, after he has read our history itself and the tables of invention, that there is some uncertainty, if not actual falsehood, in the experiments themselves, and for that reason he will perhaps think that our discoveries rest upon false and doubtful foundations and principles. But this is nothing; such things are bound to happen at the beginning. It is as if in writing or printing one or two letters happen to be poorly formed or badly placed, it does not usually bother the reader very much, since they are easily corrected by the sense itself. So men should realise that many experiments in natural history may be wrongly believed and accepted which a little while later are easily deleted and excluded from the discovery of causes and axioms. It is true however that if many mistakes are frequently and repeatedly made in natural history and experiments, they cannot be corrected or emended by any successful exercise of intelligence or art. If then there remain any faults or errors in particulars in our natural history, which has been so diligently, strictly and even religiously examined and collected, what may one say of the usual natural history, which in comparison with ours is so careless and easy-going, or about the philosophy and sciences built upon such sands (sinking sands, I would say)? Thus what we have said should bother no one.

CXIX

In our history and experiments there will also be many things which are trivial and common, many that are low and illiberal, many that are excessively subtle, merely speculative and apparently useless: the kind of thing that might turn people off and alienate their support.

Men should recognise that, with regard to ordinary things, they are very liable simply to relate and adapt the causes of rare events to things that happen frequently, but not to investigate the causes of the things that happen frequently themselves; they take them as given and admitted.

And so they do not look for the causes of weight, of the rotation of the heavenly bodies, of heat, cold, light, hard, soft, rare, dense, liquid, solid, animate, inanimate, similar, dissimilar, or of the organic; they do however argue and give judgements about other things which do not occur so frequently and are not so familiar.

But we who know well enough that no judgement can be made of rare or remarkable things, much less new things brought to light, without investigation and discovery of the causes of common things and the causes of their causes, we are necessarily forced to admit the commonest things into our history. Furthermore, we find that the greatest obstacle to the progress of philosophy has been that familiar things of frequent occurrence do not arrest and hold men's contemplation; they are barely noticed in passing, and no inquiry made as to their causes. We more often need to pay attention to known things than to get information about unknown things.

CXX

As for mean or even foul objects for which (as Pliny says) we have to apologise, they have to be admitted to natural history no less than the most elegant and valuable objects. Natural history is not polluted by them: the sun enters palaces and drains indifferently, and is not polluted. We are not building or dedicating a Capitol or a Pyramid to men's pride, but are laying the foundations in the human intellect of a holy temple on the model of the world. And so we follow the model. For whatever is worthy of being is worthy also of knowledge, which is the image of being. And mean things exist as well as elegant things. Moreover, just as the best perfumes are sometimes made from smelly things like musk and civet, so excellent light and information emerge from mean and dirty things. Enough of this; such delicacy is altogether childish and effeminate.

CXXI

But we must certainly take a more careful look at the objection that to the common understanding, or any understanding accustomed to present things, much in our history will seem to have a kind of curious and useless subtlety. This needs to be discussed before anything, and has been discussed; and this is the point: that now as we begin, and for some time to come, we are looking only for experiments that are enlightening, not productive experiments. Our model is God's creation, as we have often said, which on the first day brought forth only light, and devoted the whole day only to this, and made no work of matter on that day.

So anyone who thinks such things are useless thinks like one who believes that light is useless because it is not a solid or material thing. And in truth it must be said that a careful, well-sifted knowledge of simple natures is like light, which affords access to the universal secrets of effects, and has a kind of power to grasp and bring with it whole legions and squadrons of effects and the sources of the most remarkable axioms; but in itself it is not very useful. Just so the letters of the alphabet by themselves and separately from one another mean nothing and are no use, and yet are like the first matter for the composition and equipping of all discourse. Seeds of things are strong in power, but (except in their own process) quite useless. Dispersed rays of light itself impart no benefit unless they come together.

But if anyone is offended by speculative subtleties, what should be said of the scholastics and their massive indulgence in subtleties? Their subtleties were spent on words or (which comes to the same thing) on common notions, not on things or on nature; they were quite useless in their origin and in their consequences; and were not the sort of subtleties which have no usefulness for the present but infinite utility in the consequence, as are these of which we speak. Men should take it for certain that all subtlety in disputation and reflection which is brought to bear only after axioms have been discovered is tardy and too late. The true and proper time for subtlety, or at least the best time, is in weighing experience and in forming axioms from it. The other kind of subtlety grabs and touches nature, but never grasps or catches it. The common saying about opportunity or fortune is surely very true when applied to nature: 'it is long-haired at the front, but bald at the back'.[58]

A last remark on the contempt in natural history for things that are common or mean, or too subtle and useless in their beginning: the retort of the little woman to the pompous prince ought to be taken as an oracle; he had refused to hear her petition as too petty for him and unworthy of his majesty: 'Cease then to be king', she said.[59] For it is quite certain that empire over nature can be neither gained nor exercised if one is unwilling to attend to such things because they are too petty and too small.

CXXII

It will[60] also be objected that it is an incredibly brutal proceeding to do away with all sciences and all authors instantaneously in one sudden attack, and not to take anything from the ancients for our help and support, but to rely on our own strength.

But we know that if we had been willing to act with less than complete good honesty, it would not have been difficult for us to attribute our proposals either to the ancient centuries before the times of the Greeks (when perhaps the sciences of nature were flourishing more fully though in deeper silence, without the benefit of Greek pipes and trumpets) or even (for part of it) to some of the Greeks themselves, and to seek the assurance and honour that comes from that; like *parvenus* devising a fabricated nobility for themselves from some ancient lineage by means of genealogies. But we rely on the evidence of things, and reject even the suspicion of fiction and imposture. We do not think that it is any more relevant to the present subject whether the discoveries to come were once known to the ancients, and have been dying and recurring in the revolutions of things through the centuries, than it should matter to men whether the New World is the famous island Atlantis which the ancient world knew or a new land now discovered for the first time. For the discovery of things is to be taken from the light of nature, not recovered from the shadows of antiquity.

As for my general criticism of past sciences, that is surely both more

---

[58] Phaedrus, *Fables*, v.8; also Dionysius Cato, *Moral Distichs*, ii.26.
[59] Told of Philip II of Macedon by Plutarch, *Sayings of Kings and Commanders*, 179C.
[60] Reading *occurret* for *occurrit*.

plausible, on a true view, and more modest than a partial criticism. For if the mistakes had not been rooted in first notions, true discoveries would have been bound to correct erroneous discoveries. But when there have been fundamental mistakes, such that men have missed and ignored things rather than made a bad or incorrect judgement of them, it is not at all surprising that they have not got what they did not attempt, and have not reached a goal which they had not set or fixed, and have not completed a race which they had neither entered nor run.

With regard to the insolence of the thing: certainly if someone claims for himself that he can draw a straighter line or a more perfect circle than anyone else by steadiness of hand and sharpness of eye, he is obviously inviting a contest of abilities. But if anyone asserts that he can draw a straighter line or a more perfect circle with the aid of a rule or pair of compasses than anyone else can with his unaided eye and hand, he is surely not bragging at all. And what we say applies not only to our first, preliminary effort, but is also applicable to those who devote themselves to this subject in the future. For our method of discovery in the sciences more or less equalises intellects, and leaves little opportunity for superiority, since it achieves everything by most certain rules and forms of proof. Thus our present work (as we have often said) is rather due to a kind of good luck than to ability, and is a birth of time rather than of intellect. For there is certainly an element of chance in men's thoughts no less than in their works and deeds.

CXXIII

Therefore we must apply to ourselves the old story, especially as it cuts to the heart of the matter: 'when one man drinks water and the other wine, it is impossible for them to think alike'. All other men, ancient and modern alike, have in the sciences drunk a simple drink like water, either running spontaneously from the understanding, or drawn up by dialectic as by pulleys from a well. But we drink and make our toasts in a liquor which is made[61] from great numbers of grapes, ripe grapes, ready for the vintage, gathered and cut from the stem in selected bunches, then crushed in the winepress, then refined and strained in a vessel. And so it is no surprise if we do not agree with other people.

[61] Reading *confectum* for *confectam*.

CXXIV

Here is another objection that will certainly come up: that (despite our criticisms of others) we ourselves have not first declared the true and best goal or purpose of the sciences. For the contemplation of truth is worthier and higher than any utility or power in effects; but the long and anxious time spent in experience and matter and in the ebb and flow of particular things keeps the mind fixed on the ground, or rather sinks it in a Tartarus of confusion and turmoil, and bars and obstructs its way to the serenity and tranquillity of detached wisdom (a much more godlike condition). We willingly assent to this argument; it is precisely this thing which they hint and find preferable which we are chiefly and above all engaged on. For we are laying the foundations in the human understanding of a true model of the world, as it is and not as any man's own reason tells him it is. But this can be done only by performing a most careful dissection and anatomy of the world. We declare that the inept models of the world (like imitations by apes), which men's fancies have constructed in philosophies, have to be smashed. And so men should be aware (as we said above)[62] how great is the distance between the *illusions* of men's minds and the ideas of God's mind. The former are simply fanciful abstractions; the latter are the true marks of the Creator on his creatures as they are impressed and printed on matter in true and meticulous lines. Therefore truth and usefulness are (in this kind) the very same things,[63] and the works themselves are of greater value as pledges of truth than for the benefits they bring to human life.

CXXV

It will also perhaps be objected that we are doing something that has already been done, that the ancients themselves went the same way as we do. And so one will think it likely that after so much effort and commotion, we too will settle at last on one of the philosophies which prevailed among the ancients. For they too in commencing their reflections laid in an immense stock of examples and particulars, and arranged them in treatises by section and title, and built their philosophies and systems from them, and afterwards, having got some information on the subject, gave their

[62] 1.23.
[63] *ipsissimae res*: for the sentiment, cf. 11.4.

judgements and added occasional examples to win credence and illustrate their teaching, but thought it a waste of time and trouble to publish their notes on particulars, and their notebooks and treatises; and thus did what men do in building, namely after completion of the building, remove the scaffolding and ladders from sight. That is certainly the process one should think occurred. But one will easily meet this objection (or rather scruple) if one has not completely forgotten what we said above. For we too[64] admit there was a form of inquiry and discovery among the ancients, and their writings make clear what it was. It was simply that from some examples and particulars (with the addition of common notions, and perhaps a dose of the most popular received opinions) they leapt to the most general conclusions or principles of the sciences, by whose fixed and immovable truth they might deduce and demonstrate lesser conclusions by intermediate steps; and from these they formed their system. Then finally if new particulars and examples were proposed and adduced which ran contrary to their views, they cleverly brought them into line by means of distinctions or explanations of their own rules, or in the last analysis got rid of them altogether by making exceptions; but they laboriously and obstinately adapted to their principles the cases of particular things which were not contrary. But that was not natural history and experience as it should have been (far from it), and their hasty flight to the most general principles destroyed everything.

CXXVI

Here is another objection: that in our hesitation to make pronouncements and to lay down fixed principles until we duly arrive by way of intermediate steps at the most general principles, we maintain a kind of suspension of judgement, and bring the thing down to *Lack of Conviction* (*Acatalepsia*). But what we have in mind and propose is not *Acatalepsia* but *Eucatalepsia* (*Sound Conviction*): for we do not detract from the senses, but assist them; we do not discredit the understanding, but regulate it. And it is better to know as much as we need to know, and yet think that we do not know everything, than to think that we know everything, and yet know none of the things which we need to know.

---

[64] Rejecting Ellis's emendation *profitentur* and retaining the 1620 reading *profitemur*.

CXXVII

It may also be doubted (rather than objected) whether we are speaking of perfecting only Natural Philosophy by our method or also the other sciences, Logic, Ethics and Politics. We certainly mean all that we have said to apply to all of them; and just as common logic, which governs things by means of the syllogism, is applicable not only to the natural sciences but to all the sciences, so also our science, which proceeds by *induction*, covers all. For we are making a history and tables of discovery about anger, fear, shame and so on; and also about instances of political affairs; and equally about the mental motions of memory, composition and division, of judgement and the rest, no less than of heat and cold, or light, or vegetative growth, and so on. However, since our method of *interpretation*, after a history has been collected and organised, looks not only at the motions and activities of the mind (as the common logic does), but also at the nature of things we so govern the mind that it may apply itself to the nature of things, in ways that are suitable to all things. And therefore we give many different instructions in our teaching of *interpretation* which in some degree adapt the method of discovery to the quality and condition of the subject of inquiry.

CXXVIII

But it would be wrong even to entertain a doubt about whether we desire to destroy and abolish the philosophy, the arts and the sciences which we use; on the contrary, we happily embrace their use, their cultivation and their rewards. We do not in any way discourage these traditional subjects from generating disputations, enlivening discourse and being widely applied to professional use and the benefit of civil life, and from being accepted by general agreement as a kind of currency. Furthermore, we freely admit that our new proposals will not be very useful for those purposes, since there is no way that they can be brought down to the common understanding, except through their results and effects. But our published writings (and especially the books *On the Advancement of Learning*) testify how sincerely we mean what we say of our affection and goodwill towards the accepted sciences. And so we shall not try further to convince with words. In the meantime we give this constant and explicit warning: no great progress can be made in the doctrines and thinking of the sciences, nor can they be applied to a wide range of works, by the methods commonly in use.

CXXIX

It remains to say a few things about the excellence of the Purpose. If we had said these things before, they would have seemed like mere wishes, but now that hope has been given, and unwarranted prejudices removed, they will perhaps have more weight. And if we had completed and quite finished the whole thing, if we were not inviting others to play a part from now on and take a share in our labours, then too we would have refrained from words of this kind, in case they should be taken as a proclamation of our own merit. But since we have to excite the industry of others and stir their hearts and set them on fire, it is appropriate to recall certain things to men's minds.

First therefore, the introduction of remarkable discoveries holds by far the first place among human actions; as the ancients judged. For they ascribed divine honours to discoverers of things; but to those who had made great achievements in political matters (such as founders of cities and empires, legislators, liberators of their countries from long-standing evils, conquerors of tyrants and so on) they decreed only the honours of heroes. And anyone who duly compares them will find this judgement of antiquity correct. For the benefits of discoveries may extend to the whole human race, political benefits only to specific areas; and political benefits last no more than a few years, the benefits of discoveries for virtually all time. The improvement of a political condition usually entails violence and disturbance; but discoveries make men happy, and bring benefit without hurt or sorrow to anyone.

Again, discoveries are like new creations, and imitations of divine works; as the poet well said:

Athens, of glorious name, was once the first to give fruitful crops to men in their misery, and RECREATED their life, and made them laws.[65]

And it seems worthy of note in Solomon, that though he abounded in power, gold, magnificent works, courtiers, servants, in naval power too, and

---

[65] Lucretius, *On the Nature of Things*, VI.1-3. In the edition of Cyril Bailey (Oxford Classical Texts, 2nd edn, Oxford, 1922) the lines are printed as 'Primae frugiparos fetus mortalibus aegris/ dididerunt quondam praeclaro nomine Athenae/et recreaverunt vitam legesque rogarunt.'

the fame of his name and unparalleled human admiration, yet he selected none of these things as his glory, but declared as follows: 'the glory of God is to conceal a thing; the glory of a king is to find out a thing'.[66]

Again (if you please), let anyone reflect how great is the difference between the life of men in any of the most civilised provinces of Europe and in the most savage and barbarous region of New India; and he will judge that they differ so much that deservedly it may be said that 'man is a God to man',[67] not only for help and benefit, but also in the contrast between their conditions. And this is due not to soil, climate or bodily qualities, but to Arts.

Again, it helps to notice the force, power and consequences of discoveries, which appear at their clearest in three things that were unknown to antiquity, and whose origins, though recent, are obscure and unsung: namely the art of printing, gunpowder and the nautical compass. In fact these three things have changed the face and condition of things all over the globe: the first in literature; the second in the art of war; the third in navigation; and innumerable changes have followed; so that no empire or sect or star seems to have exercised a greater power and influence on human affairs than those mechanical things.

And it would not be irrelevant to distinguish three kinds and degrees of human ambition. The first is the ambition of those who are greedy to increase their personal power in their own country; which is common and base. The second is the ambition of those who strive to extend the power and empire of their country among the human race; this surely has more dignity, but no less greed. But if anyone attempts to renew and extend the power and empire of the human race itself over the universe of things, his ambition (if it should so be called) is without a doubt both more sensible and more majestic than the others'. And the empire of man over things lies solely in the arts and sciences. For one does not have empire over nature except by obeying her.

Besides, if the usefulness of any one particular discovery has moved men to regard anyone who could confer such a benefit on the whole human race as more than a man, how much nobler will it appear to make a discovery which may speedily lead to the discovery of all other things? And yet (simply to tell the truth) just as we owe much gratitude to light, because we in turn can see by it to find our way, practise the arts, read and recognise

---

[66] Proverbs 25:2.
[67] A saying attributed to Caecilius Comicus.

each other, and yet the actual seeing of light is a more excellent and finer thing than its many uses, so surely the very contemplation of things as they are, without superstition or deceit, error or confusion, is more valuable in itself than all the fruits of discoveries.

Finally, if anyone objects that the sciences and arts have been perverted to evil and luxury and such like, the objection should convince no one. The same may be said of all earthly goods, intelligence, courage, strength, beauty, wealth, the light itself and all the rest. Just let man recover the right over nature which belongs to him by God's gift, and give it scope; right reason and sound religion will govern its use.

CXXX

And now it is time to lay out the actual art of Interpreting Nature. Though we believe that what we teach here is what is truest and most useful, still we do not say that it is absolutely essential (as if nothing could be done without it) or even totally complete. For it is our opinion that men could hit upon our form of interpretation simply by their own native force of intelligence, without any other art, if they had available a good history of nature and experience, and worked carefully on it, and were able to give themselves two commands: one, to lay aside received opinions and notions; the other, to restrain their minds for the time being from the most general principles and the next most general. For *interpretation* is the true and natural work of the mind once the obstacles are removed; but still everything will certainly be more in readiness because of our instructions, and much more secure.

Yet we are not claiming that nothing could be added to them. On the contrary, we who look at the mind not only in its own native ability, but also in its union with things, must take the position that the art of discovery may improve with discoveries.

APHORISMS
ON
THE INTERPRETATION OF NATURE
OR ON
THE KINGDOM OF MAN
[BOOK II]

Aphorism I

The task and purpose of human Power is to generate and superinduce on a given body a new nature or new natures. The task and purpose of human Science is to find for a given nature its Form, or true difference, or causative nature or the source of its coming-to-be (these are the words we have that come closest to describing the thing). Subordinate to these primary tasks are two other tasks which are secondary and of less importance: to the first is subordinate the transformation of concrete bodies from one thing into another within the bounds of the *Possible*; to the latter is subordinate the discovery, in every generation and motion, of the continuous *hidden process* from the manifest Efficient cause and the observable matter to the acquired Form; and similarly, the discovery, in bodies at rest and not in motion, of the latent structure.

II

The sorry state of current human knowledge is clear even from common expressions. It is right to lay down: 'to know truly is to know by causes'. It is also not bad to distinguish four causes: Material, Formal, Efficient and Final. But of these the Final is a long way from being useful; in fact it actually distorts the sciences except in the case of human actions. Discovery of Form is regarded as hopeless. And the Efficient and Material causes (as they are commonly sought and accepted, i.e. in themselves and apart from the *latent process* which leads to the Form) are perfunctory, superficial things, of almost no value for true, active knowledge. Nor have we forgotten that earlier we criticised and corrected the error of the human mind in assigning to Forms the principal role in being.[1] For though nothing exists in nature except individual bodies which exhibit pure individual acts in accordance with law, in philosophical doctrine, that law itself, and the investigation, discovery and explanation of it, are taken as the foundation both of knowing and doing. It is this *law* and its *clauses*[2] which we understand by the term Forms, especially as this word has become established and is in common use.

III

He who knows the cause of a nature (as of white or of heat) only in certain subjects has an imperfect Knowledge of it; and he who can produce an effect only on some of the susceptible materials has a Power which is equally imperfect. And he who knows only the Efficient and Material causes (causes which are variable, and merely vehicles and capable of conveying forms in some things only) may achieve new discoveries in material which is fairly similar and previously prepared, but does not touch the deeply rooted ends of things. But he who knows forms comprehends the unity of nature in very different materials. And so he can uncover and bring forth things which have never been achieved, such as neither the vicissitudes of nature nor experimental efforts nor even chance have ever brought into being and which were unlikely ever to enter men's minds. Hence true Thought and free Operation result from the discovery of Forms.

IV

Although the road to human knowledge and the road to human power are very close and almost the same, yet because of the destructive and inveterate habit of losing oneself in abstraction, it is altogether safer to raise the sciences from the beginning on foundations which have an active tendency, and let the active tendency itself mark and set bounds to the contemplative part. And therefore when we think about generating and superinducing a nature on a given body, we must consider what sort of instruction and what

---

[1] *primas essentiae*: cf. 1.51 and 1.65.

[2] Bacon seems to have in mind the analogy of statute law, which was structured as a single (very long) sentence with paragraph-shaped clauses.

sort of direction or guidance one would most want; and we should do it in simple, not abstruse, language.

For example: if one wants to superinduce on silver the tawny colour of gold, or an increase of weight (with respect for the laws of the substance), or transparency on non-transparent stone, or strength on glass, or the ability to grow on something which is not vegetable, one must consider (I say) what sort of instruction or guidance a person would most wish to be given. And in the first place, he will certainly want to be shown something which would not fail in effect or disappoint in experiment. Secondly, he will desire to be prescribed something which would not force and confine him to certain ways and means of operating. For perhaps he will not have these particular means, and not have the opportunity of easily getting and procuring them. If there are other means and other ways (apart from this instruction) of producing such a nature, perhaps they will be within the power of the operator, but he will be prevented from using them because his instructions are too narrow, and he will get no results. Thirdly, he will want to be shown something which is not as difficult as the operation which he is investigating, but which comes closer to practice.

This then will have to be our declaration on the true and perfect precept of operation: *it should be certain, free and favourable to, or tending towards, action.* And this is the same as the discovery of true Form. For the form of a nature is such that if it is there, the given nature inevitably follows. Hence it is always present when the nature is present; it universally affirms it, and is in the whole of it. The same form is such that when it is taken away, the given nature inevitably disappears. And therefore it is always absent when that nature is absent, and its absence always implies the absence of that nature, and it exists only in that nature. Finally, a true form is such that it derives a given nature from the source of an essence which exists in several subjects, and which is better known to nature (as they say)[3] than the Form itself. And so our declaration and precept about the true and perfect axiom of knowledge is this: *find another nature that is convertible with a given nature, and yet is a limitation of a better-known nature,* as of *a true genus.* These two pronouncements, the active and the contemplative, are one and the same; and what is most useful in operating is truest in knowing.

---

[3] *notior naturae,* expressed also by Bacon as *natura notior,* and referring to what is more general. See, for example, 1.22.

V

The precept or axiom of the transformation of bodies is of two kinds. The first looks at the body as a company or combination of simple natures. For example, the following things are all found together in gold; it is tawny-coloured; it is heavy with a certain weight; it is malleable or ductile to a certain degree; it is not volatile, and loses none of its quantity in fire; it melts with a certain fluidity; it is separated and dissolved in certain ways; and so on for the rest of the natures which are found together in gold. Thus this kind of axiom derives the object from the forms of simple natures. For he who knows the forms and methods of superinducing tawny colour, weight, ductility, stability, melting, solution and so on, and their degrees and manners, will take pains to try to unite them in some body, and from this follows the transformation into gold. This kind of operation is a primary action. For it is the same method to generate some one simple nature as several, except that there is more constraint and restriction in operating if several are required, because of the difficulty of uniting so many natures, which are not easily brought together except by the common, ordinary ways of nature. It must in any case be said however that this mode of operation (which looks at simple natures, albeit in a compound body) proceeds from what is constant, eternal and universal in nature, and affords vast opportunities to human power, such as human thought (as things are now) can scarcely conceive or imagine.

But the second kind of axiom (which depends on the discovery of the *latent process*) does not proceed by simple natures, but by compound bodies as they are found in nature in the ordinary course of things. This is so, for example, in the case where one is investigating the origins, means and process by which gold or any other metal or stone is generated from their base substances or elements to the perfect mineral; or similarly the process by which plants are generated, from the first solidifying of the sap in the soil, or from seeds, up to the formed plant, with constant succession of motion, and with diverse yet continuous efforts of nature; likewise, of the orderly progress of the generation of animals from conception to birth; and similarly of other bodies.

For this investigation looks not only at the generation of bodies, but also at other movements and workings of nature. For example, it looks at the case where the inquiry is about the universal process and continuous action of nutrition, from the first ingestion of food to its perfect assimilation; or similarly, about voluntary motion in animals, from the first impression on the imagination and the continued efforts of the spirit right up to the flexing and moving of the limbs; or [about the process] from the unfolding of the tongue, the lips and the other organs to the uttering of articulate sounds. For these too are concerned with compound natures, or natures which are joint members of a structure; and they have regard to special and particular habits of nature, not the fundamental and common laws which constitute Forms. Nevertheless, one must fully admit that this method looks easier and more available, and offers more hope, than the primary one.

Likewise the operative function which corresponds to this contemplative function extends the operation and moves it on from things ordinarily found in nature to things close to them or not too remote. But deeper, radical operations on nature depend altogether on primary axioms. Furthermore, where man has not been granted the right to operate but only to know, as in the case of celestial objects (for man is not permitted to operate on celestial things or to alter or transform them), still investigation of the fact itself or of the truth of the matter, no less than knowledge of causes and agreements, takes one back to the primary, universal axioms about simple natures (as about the nature of spontaneous rotation, about attraction or magnetic force, and about several other things which are more common than the celestial things themselves). One cannot expect to settle the question whether the earth or heaven really turns in daily motion without first understanding the nature of spontaneous rotation.

VI

The *latent process* of which we speak is a very different thing from anything that will readily occur to men's minds (given present preconceptions). For we do not mean actual measures, signs or stages of a process which are visible in bodies, but a wholly continuous process which for the most part escapes the senses.

Example: in every case of generation and transformation of a body we have to ask what is lost and disappears; what remains and what accrues; what expands and what contracts; what is combined, what is separated; what is continuous, what interrupted; what impels, what obstructs; what prevails, what submits; and several other questions.

Nor are these questions to be asked only in cases of generation and transformation of bodies. In the case of all other modifications and motions we must similarly ask what precedes, what succeeds; what is more pressing, what more relaxed; what furnishes motion and what guides it; and so on. All these things are unknown and unbroached by the sciences (which are currently practised by the dullest and most unsuitable persons). For since every natural action is transacted by means of the smallest particles, or at least by things too small to make an impression on the senses, no one should expect to master or modify nature without taking the appropriate means to grasp and take note of them.

VII

Similarly, the investigation and discovery of the *latent structure* in bodies is a new thing, no less than the discovery of *latent process* and form. We are clearly still hovering about the anterooms of nature and are not achieving entrance to her inner chambers. But no one can endow a given body with a new nature or successfully and appropriately transmute it into a new body without possessing a good knowledge of modification or transformation of body. He will find himself using useless methods, or at least difficult and cumbersome methods unsuitable for the nature of the body on which he is working. Thus here too the road needs to be opened and constructed.

It is surely right and useful to spend effort on the anatomy of organic bodies (as of man and of the animals), and it seems to be a subtle thing and a good search of nature. This kind of anatomy is perceptible and open to the senses and appropriate only in the case of organic bodies. But it is an obvious and easily available thing compared with the true anatomy of latent structure in bodies regarded as similar; especially in things of the same species and their parts, as iron and stone; and in the similar parts of a plant or animal; as the root, the leaf, flower, flesh, blood, bone etc. But even in this case human industry has not wholly failed; this is actually the tendency of distillations and other methods of solution, that the dissimilarity of a compound should appear through the gathering together of homogeneous parts. And this is useful, and helps towards what we are looking for, though often enough the thing is deceptive, because several natures are assigned and attributed to the separated substance as if they had subsisted previously in the compound, whereas in truth fire and heat and the other agents of opening it up give and superinduce them for the first time. But this too is a small part of the work of discovering the true structure in the compound; this structure is a much more subtle and precise thing, and is rather obscured by the effects of fire than revealed and illuminated.

Therefore separation and dissolution of bodies is certainly not to be achieved through fire, but by reason and true induction, with auxiliary experiments; and by comparison with other bodies, and the reduction to simple natures and their forms which assemble and unite in the compound; one must make a clean break with *Vulcan*, and move to the side of *Minerva*, if one wants to bring into the light the true textures and structures of bodies (and every hidden and, as they say, specific property and power in things depends upon this; and also therefore every rule for effective modification and transformation derives from it).

Example: one must ask of every body how much spirit there is in it, and how much tangible essence; and of the spirit itself ask whether it is abundant and swelling, or weak and sparse; thinner or denser; tending to air or fire; sharp or sluggish; feeble or robust; advancing or retreating; broken or continuous; at home or at odds with the surrounding environment, etc. Likewise the tangible essence (which allows as many differences as spirit), with its hairs and fibres and textures of every kind, is subject to the same inquiry, and so is the distribution of the spirit through the bodily mass, and its pores, passages, veins and cells, and the rudiments or first attempts at organic body. True, clear light is shed on these too, and thus on every discovery of *latent structure*, by the primary axioms, which surely dispel all darkness and every subtlety.

VIII

However, we will not end up with the atom, which presupposes a vacuum and unmoving matter (both of which are false), but to true particles as they are found to be. But there is no reason why anyone should shy away from this subtlety as inexplicable; on the contrary, the more the inquiry moves towards simple natures, the more all things will be in a plain, transparent light; as the procedure passes from the multiple to the simple, from the incommensurable to the commensurable, from the random to the calculable, and from the infinite and undefined to the definite and certain; as it is with the letters in writing and the notes in chords. Natural inquiry succeeds best when the physical ends in the mathematical. And no one should be afraid of multiplicity or of fractions. For in numerical calculations one would as easily posit or think of a thousand as of one, or of a thousandth part as easily as of a whole.

IX

A true division of philosophy and the sciences arises from the two kinds of axioms which have been given above, if we translate the normal words (which come closest to indicating the thing) into our own terms. The inquiry after *forms*, which are (at least by reason and their law) eternal and unmoving, would constitute *metaphysics*; the inquiry after the *efficient* and *material causes, the latent process* and *latent structure* (all of which are concerned with the common and ordinary course of nature, not the fundamental, eternal laws) would constitute *physics*; subordinate to these in the same manner are two practical arts: *mechanics* to *physics*; and *magic* to *metaphysics* (in its reformed sense), because of its broad ways and superior command over nature.

X

Having laid down the scope of our teaching, we proceed to precepts; and in the least awkward and unnatural order. Directions for *the interpretation of nature* comprehend in general terms two parts: the first for drawing axioms from experience; the second on deducing or deriving new experiments from axioms. The former is divided three ways, i.e. into three kinds of service: service to the senses, service to the memory and service to the mind or reason.

First we must compile a good, adequate *natural* and *experimental* history. This is the foundation of the matter. We must not invent or imagine what nature does or suffers; we must discover it.

A *natural* and *experimental history* is so diverse and disconnected that it confounds and confuses the understanding unless it is stopped short, and presented in an appropriate order. So *tables* must be drawn up and a *co-ordination of instances* made, in such a way and with such organisation that the mind may be able to act upon them.

Even with these, the mind, left to itself and moving of its own accord, is incompetent and unequal to the formation of axioms unless it is governed and directed. And therefore, in the third place, a true and proper induction must be supplied, which is the very *key of interpretation*. And one must begin at the end and move backwards to the rest.

XI

The investigation of forms proceeds as follows: first, for any given nature one must make a *presentation*[4] to the intellect of all known *instances* which meet in the same nature, however disparate the materials may be. A collection of this kind has to be made historically, without premature reflection or any great subtlety. Here is an example in the inquiry into the form of heat.

[Table 1]
Instances meeting in the nature of heat

1. the sun's rays, especially in summer and at noon
2. the sun's rays reflected and concentrated, as between mountains or through walls, and particularly in burning glasses
3. flaming meteors
4. lightning that sets fires
5. eruptions of flame from hollows in mountains etc.
6. any flame
7. solids on fire
8. natural hot baths
9. heated or boiling liquids
10. steam and hot smoke, and air itself, which is capable of a powerful, furious heat if compressed, as in reverse furnaces[5]
11. some spells of weather which are clear and bright through the actual constitution of the air without regard to the time of the year
12. air shut up underground in some caverns, especially in winter
13. all fibrous fabrics, such as wool, animal hides and plumage, have some warmth
14. all bodies, solid and liquid, thick and thin (like the air itself) brought close to a fire for a time

---

[4] *Comparentia* is a legal term which refers to the 'presentment' of the defendant or of documents in court. 'Presentation of instances' is intended to preserve something of the legal analogy.
[5] '"Reverbatories" are furnaces constructed with two chambers; an outer one, which has no chimney, but has a passage connecting it with an inner one which has a chimney' (Kitchin).

15. sparks from flint and steel sharply struck
16. any body forcefully rubbed, as stone, wood, cloth etc.; so that yoke-beams and wheel axles sometimes catch fire; and the Western Indians have a way of making fire by rubbing
17. green, wet plants confined and compressed, like roses, peas in baskets; so that hay often catches fire if it is stored wet
18. quicklime sprinkled with water
19. iron as it is dissolved by *aqua fortis* in a glass without any use of fire; and likewise tin etc., but not so intensely
20. animals, especially internally, where they are constantly hot, though in insects the heat is not perceptible to the touch because they are so small
21. horse shit, and similar animal excrement, when fresh
22. strong oil of sulphur and of vitriol give the effect of heat in scorching linen
23. oil of marjoram and suchlike give the effect of heat when they burn the gums
24. strong distilled spirit of wine gives the effect of heat; so that if the white of an egg is dipped in it, it solidifies and goes white, almost like a cooked eggwhite; and bread dipped in it dries up and goes crusty like toast
25. spices and hot plants, like *dracunculus*, old nasturtium[6] etc., though they are not hot to the hand (neither whole nor powdered), but with a little bit of chewing are felt as hot to the tongue and palate, and almost burning
26. strong vinegar and all acids cause a pain which is not much different from the pain of heat if applied to a skinless part of the body, like the eye or the tongue, or any other part where there is a wound and the skin has been wounded and the skin has been broken
27. even sharp, intense cold induces a kind of burning sensation: for 'the penetrating cold of the North Wind burns'[7]
28. other things.

We call this *the table of existence and presence.*

---

[6] Apparently watercress.
[7] An adaptation of Virgil, *Georgics*, I.92-3: 'ne tenues pluviae rapidive potentia solis/acrior aut Boreae penetrabile frigus adurat'.

XII

Secondly, we must *make a presentation*[8] to the intellect of *instances* which are devoid of a given nature; because (as has been said) the form ought no less to be absent when a given nature is absent than present when it is present. But this would be infinite if we took them all.

And therefore we should attach *negatives* to our *affirmatives*, and investigate absences only in subjects which are closely related to others in which a given nature exists and appears. This we have chosen to call the *table of divergence*, or of closely *related absences*.

[Table 2]
Closely related instances which are devoid of the nature of heat

1. *the first negative or attached instance to the first affirmative instance.*[9] The moon's rays and those of the stars and comets are not found to be hot to the touch; moreover, the sharpest frosts are normally observed at the full moon. But the larger fixed stars are thought to increase and intensify the heat of the sun when it goes under them or approaches them; as happens when the sun is in Leo and in the dog days.

2. *negative to the second affirmative instance.* The sun's rays do not give off heat in the middle region of the air (as they call it); the common explanation of this is quite good, that that region does not come close enough to the body of the sun, from which the rays emanate, nor to the earth, by which they are reflected. This is clear from the tops of mountains (unless they are particularly high) where the snows are perpetual. On the other hand some travellers have remarked that at the summit of the Peak of Tenerife[10] and also on the Peruvian Andes, the actual peaks of the mountains are destitute of snow; the snow lies only on the lower slopes. And also on the actual summits the air is observed not to be cold but thin and sharp; so that in the Andes its excessive sharpness stings and hurts the eyes, and also stings the mouth of the stomach and causes vomiting. It was also noticed by writers in antiquity that the air at the top of Olympus[11] was so

---

[8] See note on *comparentia* at II.11 above.

[9] This and the similar subtitles in each of the 32 Instances is printed as a marginal note in the Latin text.

[10] In the Canary Islands.

[11] The legendary home of the gods in ancient Greece. Hence the reference to the altar of Jupiter (or Zeus) below.

thin that those who made the ascent had to take sponges soaked in vinegar and water and apply them from time to time to mouth and nostrils, because the thinness of the air made it inadequate for breathing. It was also said of that peak that it was so calm and undisturbed by rain and snow that letters traced with their fingers by the celebrants in the ashes on the altar of Jupiter remained there undisturbed till the following year. And even today those who ascend to the top of the Peak of Tenerife do so at night and not in the day; and soon after sunrise are advised and prompted by their guides to make their descent quickly because of the danger (it seems) from the thinness of the air that it will interfere with their breathing and choke them.

3. *to the second.* In the regions near the polar circles, the reflection of the sun's rays is found to be very weak and unproductive of heat. And so the Dutch who wintered in Novaya Zemlya,[12] waiting for their ship to be freed and released by the pack ice which was holding it fast, gave up hope about the beginning of July, and had to take to the longboats. So the direct rays of the sun seem to have little power even on flat terrain; nor do reflected rays, unless they are multiplied and combined, as happens when the sun approaches the perpendicular. The reason is that at that time the incidence of the rays forms quite acute angles, so that their lines are closer together; by contrast, when the inclinations are high, the angles are very obtuse, and consequently the lines of the rays further apart. However, one should note that there may be many ways in which the sun's rays may work, as well as from the nature of heat, which are not suited to our touch, so that they do not cause heat for us but do produce the effects of heat for some other bodies.

4. *to the second.* Try this experiment: take a lens made the opposite way from a burning-glass, and place it between the hand and the sun's rays; and observe whether it diminishes the heat of the sun as a burning-glass increases and intensifies it. For it is clear in the case of optical rays that the images appear wider or narrower, according to the thickness of the lens at the centre and the edges respectively. The same thing should be studied with regard to heat.

5. *to the second.* Carefully try an experiment whether by means of the strongest and best-made burning-glasses the rays of the moon can be caught and combined to produce even the smallest degree of heat. If perhaps the degree of heat is too subtle and weak to be perceptible and

---

[12] The Dutch explorer Willem Barents died in 1597, in the incident described in the text, while seeking the North-East Passage.

observable to the touch we shall have to try the glasses[13] which indicate the hot or cold constitution of the air. Let the rays of the moon fall through the burning-glass and be cast on the top of a glass of this kind; and take note whether a depression of the water occurs due to heat.

6. *to the second.* Train a burning-glass on a hot body which is not radiant or luminous, e.g. iron and stone which is heated but not on fire, or boiling water, and so on; and note whether an increase and intensification of heat occurs, as with the sun's rays.

7. *to the second.* Train a burning-glass on an ordinary flame.

8. *to the third.* Comets (if we may regard them as a kind of meteor[14]) are not found to have a regular or obvious effect in increasing seasonal temperatures, though dry spells have often been noticed to follow them. Moreover, beams and columns and gulfs of light[15] and such things appear more often in the winter than in the summer; and especially in very intense cold spells, which are also dry spells. But forked lightning and sheet lightning and thunder rarely occur in winter; rather at the time of the greatest heats. But the so-called falling stars are commonly thought to consist of some viscous material which is shining and burning, rather than to be of a powerful fiery nature. But this needs further inquiry.

9. *to the fourth.* There is some sheet lightning which gives light but does not burn; it always occurs without thunder.

10. *to the fifth.* Outbursts and eruptions of flame are found in cold, no less than in hot, regions, e.g. in Iceland and Greenland; just as trees too are sometimes more inflammable - more pitchy and resinous - in cold than in hot regions; as is the case with e.g. the fir, the pine and so on. But there has not been enough inquiry into what sort of situation and kind of terrain such eruptions normally occur in, to enable us to append a *negative* to the *affirmative*.

11. *to the sixth.* All flame is more or less hot, and there is no *negative* attached; however, they do say that the so-called *ignis fatuus*, which even sometimes settles on a wall, does not have much heat, perhaps like the flame of spirit of wine, which is gentle and weak. The flame which is found appearing around the heads and hair of boys and girls in some serious, reliable histories seems to be still weaker; it did not burn the hair at all but softly flickered around it. It is also quite certain that a kind of gleam

13 Thermometers
14 Or perhaps 'heavenly body'.
15 This seems to be a reference to the aurora borealis.

without obvious heat has appeared around a horse sweating as it travelled at night in clear weather. A few years ago a notable incident occurred which was almost taken for a miracle: a girl's girdle flashed when it was moved or rubbed a little; this may have been caused by the alum or salts with which the girdle had been soaked forming a thick coat on it which became crusted, and being broken by the rubbing. It is also certain that all sugar, whether refined (as they say) or raw, provided it is quite hard, sparkles when broken or scraped with a knife in the dark. Similarly salt seawater is sometimes found to sparkle at night when forcefully struck by oars. And in storms highly agitated sea foam gives off a flash; the Spanish call this flash *the lung of the sea*. There has not been enough investigation of how much heat is given off by the flame which sailors in the ancient world called *Castor and Pollux* and today is called *St Elmo's Fire*.

12. *to the 7th.* Everything which has been burned so that it turns to a fiery red is always hot even without flame, and no *negative* is attached to this *affirmative*. The closest thing [to a negative instance] seems to be rotten wood, which shines at night but is not found to be hot, and the rotting scales of fish, which also shine at night but are not found to be hot to the touch. Nor is the body of the glow-worm or of the fly which they call 'fire-fly' found to be hot to the touch.

13. *to the 8th.* There has not been enough investigation of the locations and nature of the earth from which hot springs flow; so no *negative* is attached.

14. *to the 9th.* The *negative* attached to hot liquids is liquid itself in its own nature. For no tangible liquid is found which is hot in its nature and constantly stays hot; rather heat is superinduced for a time only as an adventitious nature. Hence liquids that are the hottest in their power and operation, like spirit of wine, chemical oil of spices, and oil of vitriol and of sulphur, and suchlike, which quickly cause burning, are cold at first touch. And when water from hot springs is collected in a pitcher and taken away from the springs, it cools down, just like water heated by fire. Oily substances, it is true, are less cold to the touch than watery substances, as oil is less cold than water and silk less cold than linen. But this belongs to the table of Degrees of Cold.

15. *to the 10th.* Similarly the *negative* attached to hot steam is the nature of steam itself as we experience it. Emissions from oily substances, though readily inflammable, are not found to be hot unless just emitted from a hot body.

16. *to the 10th.* Just so the negative attached to hot air is the nature of air itself. For we do not experience air as hot unless it has been confined or subjected to friction or obviously heated by the sun's fire or some other hot body.

17. *to the 11th.* The *negative* attached is periods which are colder than normal at that season, which occur among us when the East or the North winds are blowing; just as the opposite kind of weather occurs when the South and West winds are blowing. A tendency to rain (especially in winter) goes along with warm weather, and frost with cold weather.

18. *to the 12th.* The attached negative instance is air confined in caves in summertime. But an altogether more thorough investigation is needed of confined air. For first there is a reasonable doubt as to the nature of air in relation to heat and cold in its own proper nature. For air obviously receives heat from the influence of the heavenly bodies; and cold perhaps by emission from the earth; and in what they call the middle region of the air, from cold fogs and snow; so that no judgement can be made of the nature of air from air which is outside and in the open, but might be made more accurately from confined air. And it also necessary for the air to be confined in a jar and in material which neither affects the air with heat or cold of its own, nor easily admits the influence of air from outside. Let the experiment be made therefore with an earthenware jar wrapped in several layers of leather to protect it from the outside air, sealing it well and keeping the air in it for three or four days; take the reading after opening the jar, either by hand or by carefully applying a thermometer.

19. *to the 13th.* Similarly, there is some doubt whether the heat in wool, skins, feathers and so on comes from a feeble heat inhering in them because they have been stripped from animals; or also because of a certain fattiness and oiliness, which is of a nature akin to warmth; or simply because air is confined and cut off, as described in the preceding paragraph. For all air cut off from contact with the outside air seems to have some warmth. So let an experiment be made on fibrous material made from flax, not on wool, feathers or silk, which are stripped from animals. Notice too that every kind of dust (which obviously traps air) is less cold than the corresponding whole bodies from which the dust came; just as we also suppose that spray of any kind (since it contains air) is less cold than the actual liquid.

20. *to the 14th.* This has no *negative* attached. For we find nothing either tangible or spirituous which does not take on heat when brought close to fire. These things do however differ from each other in that some absorb

heat quickly, like air, oil and water, while others do so more slowly, like stone and metals. But this belongs to the *Table of Degrees*.

21. *to the 15th.* There is only one *negative* attached here: notice that sparks are only struck from flint, steel or any other hard substance when minute fragments of stone or metal are struck off from the substance itself, and air subjected to friction never generates sparks of itself, as is commonly thought. Moreover the sparks themselves shoot downwards rather than up, because of the weight of the body ignited, and when extinguished turn back into a sooty substance.

22. *to the 16th.* We believe there is no *negative* attached to this instance. For we find no tangible body which does not manifestly grow warm by rubbing; so that the ancients imagined that there was no other means or virtue of heating in the heavenly bodies than from the friction of the air by rapid and intense rotation. We must ask a further question on this subject: do bodies ejected from machines (as balls from cannons) acquire some heat from the blast itself; so that they are found to be quite hot when they fall? Air in motion rather cools than heats, as in winds, bellows and the expulsion of air through pursed lips. But motion of this kind is not rapid enough to cause heat, and acts according to the whole, not by particles, so that it is no wonder if it does not generate heat.

23. *to the 17th.* A more careful inquiry needs to be made of this instance. For herbs and vegetables when green and moist seem to have some hidden heat in themselves. This heat is so slight that it is not perceptible to touch in an individual instance. But when they have been put together and confined, so that their spirit does not escape into the air but nurtures itself, then indeed a noticeable heat arises, and sometimes a fire if the material is suitable.

24. *to the 18th.* We must also make a more thorough investigation of this instance. For quicklime sprinkled with water seems to generate heat either because of the concentration of heat previously dispersed (as we said above about stored herbs), or because the fiery spirit is irritated and angered by the water, and some kind of struggle and rejection of the contrary nature takes place[16]. It will readily be apparent which of these it is if we use oil instead of water; for the oil will have the same effect as water in forming a union with the enclosed spirit, but not in irritating it. Wider experiment should also be made with the ashes and limes of different bodies as well as by dropping different liquids on them.

[16] For *antiperistasis* (rejection of the contrary nature) see II.27 (towards the end).

25. *to the 19th.* Attached to this instance is the *negative* instance of other metals which are softer and more soluble. For goldleaf dissolved into a liquid by means of *aqua regis* offers no heat to the touch in its dissolving; nor likewise does lead in *aqua fortis*, nor quicksilver either (as I recall). But silver itself causes a little heat and so does copper (as I recall), and so more obviously do tin and, particularly, iron and steel, which give off not only a fierce heat on dissolving but also violent bubbling. Therefore the heat seems to be caused by the conflict when the strong waters penetrate, pit and disintegrate the parts of the body, and the bodies themselves resist. But when the bodies easily give in, hardly any heat is generated.

26. *to the 20th.* There is no *negative* attached to the heat of animals, except of insects (as remarked) because of the small size of their bodies. For in fish, as compared with land animals, it is more a matter of degree of heat than of its absence. In vegetables and plants no degree of heat is perceptible to the touch, neither in their resin nor in uncovered pith. But in animals a great range of heat is found, both in their parts (for the amounts of heat around the heart, in the brain and around the external parts are all different), and in their occasional states, as in violent exercise and fevers.

27. *to the 21st.* There is hardly any *negative* to this instance. Even animal excrement which is not fresh has potential heat, as is seen by its fertilisation of the soil.

28. *to the 22nd and 23rd.* Liquids (whether denominated waters or oils) which have a high and intense acidity act like heat in tearing bodies apart and eventually burning them, but they are not hot to the touch of a hand at the beginning. They operate by affinity and according to the porosity of the body to which they are attached. For *aqua regis* dissolves gold, but not silver; on the other hand *aqua fortis* dissolves silver, but not gold; and neither dissolves glass. And so with the rest.

29. *to the 24th.* Make an experiment with spirit of wine on wood and also on butter, wax or pitch, to see whether it dissolves them by its heat. Instance 24 shows its power of imitating heat in producing incrustations. Let a similar experiment be made for liquefactions. Also experiment with a thermometer or calendar glass[17] shaped into a hollow bowl at the top; pour into the hollow bowl some well-distilled spirit of wine, put a lid on it to help keep its heat in; and note whether it makes the water go down by its heat.

---

[17] *vitrum graduum sive calendare.* (The term 'calendar glass' is borrowed from Ellis.)

30. *to the 25th.* Spices and herbs which are bitter to the palate, and even sharper when swallowed, feel hot. We must therefore see in what other materials they have the effect of heat. Sailors tell us that when heaps and piles of spices are suddenly opened after being long shut up, there is a danger to those who first disturb them and take them out, from fevers and inflammations of the spirit. Similarly, an experiment can be made whether the powders from spices or herbs of this kind do not dry bacon and meat hung over them like the smoke from a fire.

31. *to the 26th.* There is a biting and penetrative power both in cold things, such as vinegar and oil of vitriol, and in hot things such as oil of marjoram and suchlike. And they equally cause pain in living things, and non-living things they pull to pieces and eat away. There is no *attached* negative instance. In animate beings there is no feeling of pain without a sensation of heat.

32. *to the 27th.* Several actions of heat and cold are the same, though they work in a quite different way. For snow too seems to burn the hands of boys quite quickly; and cold keeps meat from going off no less than fire; and heat contracts bodies as cold also does. But it is more appropriate to deal with these and similar questions in the Investigation of Cold.

XIII

Thirdly, we must make a presentation to the intellect of instances in which the nature under investigation exists to a certain degree. This may be done by comparing the increase and decrease in the same subject, or by comparing different subjects with another. For the form of a thing is the very thing itself; and a thing does not differ from its form other than as apparent and actual differ, or exterior and interior, or the way it appears to us and the way it is in reality; and therefore it quite surely follows that a nature is not accepted as a true form unless it always decreases when the nature itself decreases, and likewise always increases when the nature itself increases. We have chosen to call such a table a *Table of Degrees* or *Table of Comparison*.

[Table 3]
Table of Degrees or *Comparison* on Heat

First then we will speak of things which have absolutely no degree of heat to the touch, but seem to have only a kind of potential heat, a disposition towards heat, or susceptibility to heat. Then we will move to things which are actually hot or hot to the touch, and their strengths and degrees.

1. Among solid and tangible bodies none is found that is hot in its nature originally. No stone, metal, sulphur, fossil, wood, water or animal corpse is found to be hot. The hot waters in natural baths seem to be heated by accident, whether by an underground flame or fire such as spews out of Etna and a number of other mountains, or from conflict between bodies in the way heat is produced in the solution of iron and tin. And so to the human touch, the degree of heat in inanimate objects is nil; and yet they do differ in degree of cold; for wood and metal are not equally cold. But this belongs to the *Table of Degrees on Cold*.

2. Nevertheless so far as potential heat and readiness to take fire is concerned, quite a few inanimate things are found which are extremely susceptible to heat, such as sulphur, naphtha and petroleum.

3. Things that have been hot before retain some latent relics of their former heat, as horse dung retains the heat of the animal, and lime, or perhaps ash or soot, the heat of the fire. Thus bodies buried in horse dung exude certain fluids and disintegrate, and heat is roused in lime by sprinkling water on it, as I have explained before.

4. Among vegetables no plant or part of a plant (as resin or pith) is found which is hot to human touch. Nevertheless (as said above) stored green herbs do grow warm; and some vegetables are found to be hot, others cold, to the interior touch, e.g. to the palate or stomach, or even to exterior parts after a certain time (as in the case of poultices and ointments).

5. Nothing hot to human touch is found in the parts of animals after they have died or been separated from the body. Even horse dung loses its heat unless shut in and buried. However, all dung seems to have potential heat, as in fertilising the fields. And similarly corpses of animals have a latent and potential heat of this kind; so that in cemeteries where burials occur every day, the earth acquires a kind of hidden heat which consumes a recently buried body much more quickly than fresh earth. There is a story that a kind of fine, soft cloth is found among Orientals which is made from the plumage of birds, and has an innate power to dissolve and liquefy the butter which is loosely wrapped in it.

6. Things which fertilise the fields, as dung of all kinds, chalk, sand from the sea, salt and suchlike have some inclination to heat.

7. All rotting has some traces of a weak heat in it, though not to the extent that it can be felt by touch. For things like flesh and cheese, which rot and

dissolve into little creatures, are not felt as hot to the touch; nor is rotten wood that shines at night, found to be hot to the touch. However, heat in rotting things sometimes shows itself by strong, vile smells.

8. Thus the first degree of heat, from things which are felt as hot to human touch, seems to be that of animals, which has quite a wide range of degrees. For the lowest degree (as in insects) is barely perceptible to touch; the highest degree scarcely reaches the degree of heat of the sun's rays in the hottest regions and seasons, and is not too fierce to be tolerated by the hand. And yet they say of Constantius and of some others who were of very dry constitution and bodily condition, that when they were in the grip of highly acute fevers, they almost seemed to burn the hand that touched them.

9. Animals increase in heat from movement and exercise, from wine and eating, from sex, from burning fevers and from pain.

10. At the onset of intermittent fevers, animals at first are seized by cold and shivering, but soon become exceedingly hot; as they also do right from the beginning in the case of burning and pestilential fevers.

11. We should make further investigation of the comparative heat in different animals, as in fish, quadrupeds, snakes and birds; and also by species, as lion, kite, man etc.; for in the common belief fish are quite cool internally, whereas birds are very hot, especially doves, hawks and sparrows.

12. We should make further investigation of comparative heat in the same animal, in its different organs and limbs. For milk, blood, sperm and eggs are found to be moderately warm, and less so than the exterior flesh of the animal when it is moving or agitated. Likewise no one has yet inquired into the degree of heat in the brain, stomach, heart and so on.

13. In winter and cold weather all animals are cold externally; but internally they are thought to be even warmer than usual.

14. Even in the hottest part of the world and at the hottest times of the year and the day, the heat of the heavenly bodies does not reach a sufficient level to burn or scorch the driest wood or straw or even tinder, unless it is intensified by burning-glasses; and yet it can raise a steam from damp matter.

15. The received wisdom of the astronomers makes some stars hotter and some cooler. Mars is said to be the hottest after the sun, then Jupiter, then Venus; the moon is said to be cold and Saturn coldest of all. Among the fixed stars Sirius is said to be hottest, then the Heart of the Lion, or Regulus, then the Dog-star, etc.

16. The sun has more warming power the nearer it approaches the perpendicular, or Zenith, as we should also expect of the other planets in their different degrees of heat; for example, that Jupiter is more warming for us when it lies beneath the Crab or the Lion than when it is beneath Capricorn or Aquarius.

17. We should also expect that the sun itself and the other planets have more warming power at their perigees, because of their proximity to the earth, than at their apogees. And if it should happen that in any region the sun was both at its perigee and close to the perpendicular at the same time, it would necessarily have more warming power than in a region where it was at its perigee but shining more obliquely. Hence we need to make a comparative study of the heights of the planets with regard to their closeness to the perpendicular and their obliquity, for each different region.

18. The sun, and the other planets likewise, are also thought to have more warming power when they are in proximity to the major fixed stars; as when the sun lies in Leo it is nearer to the Cor Leonis, the Cauda Leonis, the Spica Virginis and Sirius and the Dog-star, than when it lies in Cancer, where however it is more towards the perpendicular. And we have to believe that some parts of the sky give off more heat (though imperceptible to touch) because they are furnished with more stars, particularly the larger ones.

19. In general the heat of heavenly bodies is increased in three ways: viz. by perpendicularity, from propinquity or perigee, and from a constellation or company of stars.

20. In general the heat of animals and also of the heavenly rays (as they reach us) is very different from even the gentlest flame or burning objects, and also from liquids or air itself when strongly heated by fire. For the flame of spirit of wine, even in a natural, unfocused form, is still able to set fire to straw or linen or paper; which the heat of an animal or of the sun will never do without burning-glasses.

21. In flames and burning objects there are very many degrees of strength and weakness of heat, but no careful inquiry has been made about them, and so we must deal with them superficially. The flame from spirit of wine seems to be the gentlest of flames; unless perhaps *ignis fatuus* or flames or flashes from the sweat of animals are gentler. We believe that the next flame is that from light and porous plant matter, like straw, rushes and dry leaves, and the flame from hairs or feathers is not much different. Next perhaps is the flame from wood, especially the kinds of wood without much

resin or pitch, bearing in mind that the flame from less hefty sticks (which are usually bound into bundles) is smoother than that from trunks and roots of trees. This may be commonly experienced in furnaces that smelt iron, in which the fire from firewood and branches of trees is not very useful. Next comes (as we think) the flame from oil, tallow, wax and similar fatty substances which do not have much bite. The most powerful heat is found in pitch and resin, and still more in sulphur, camphor, naphtha, saltpetre and salts (after the crude matter has blown away), and in their compounds, like gunpowder, Greek fire (which is commonly called wildfire) and its different kinds, whose heat is so stubborn it is not easily extinguished by water.

22. We also think that the flame which comes from some imperfect metals is very strong and fierce. But all these things need further inquiry.

23. The flame of forked lightning seems to surpass all these flames; so that it has sometimes melted wrought iron itself into drops, which those other flames cannot do.

24. There are different degrees of heat also in bodies that have been set on fire. No careful investigation has been made of this either. We believe that the weakest is given off by burning tinder, such as we use to start a fire; and the same for the flame from the porous wood or dry cords which are used for firing cannon. Next to this is burning coal from logs and coal, and also from fired bricks and suchlike. Of fired substances we think that fired metals (iron, copper and so on) have the fiercest heat. But this too needs further investigation.

25. Some fired things are found to be far hotter than some flames. For example, fired iron is much hotter and more destructive than the flame of spirit of wine.

26. Many things too which without being on fire are simply heated by fire, like boiling water and air confined in reverse furnaces, are found to surpass in heat many kinds of flame and burning substances.

27. Motion increases heat, as one can see in the case of bellows and blasts of breath; so that the harder metals are not dissolved or melted by a dead or quiet fire but only if it is rekindled by blowing.

28. Try an experiment with burning-glasses in which (as I recall) the following happens: if a burning-glass is placed (for example) at a distance of a span[18] from a combustible object, it does not burn or consume as much

---

18 A span is nine inches.

as if it is placed at a distance of (for example) a half-span, and is slowly and by degrees withdrawn to the distance of a span. The cone and the focus of the rays are the same, but the actual motion intensifies the effect of the heat.

29. Fires occurring when a strong wind is blowing are thought to advance further against the wind than with the wind; this is because the flame leaps back with a swifter motion when the wind drops than it advances when the wind is pushing it forward.

30. A flame does not shoot out or start up unless there is an empty space in which it may move and play; except in the blasting flames of gunpowder and suchlike, where the compression and confinement of the flame intensifies its fury.

31. The anvil is much heated by the hammer; so that if an anvil were made of a thinnish sheet of metal, we would think it could be made to glow like fired iron under continued blows from a hammer; but the experiment should be tried.

32. In burning porous substances in which there is space for the fire to move, it is instantly extinguished if its motion is suppressed by strong compression, as when tinder or the burning wick of a candle or lamp or even a burning coal or lump of charcoal is snuffed out with an extinguisher or ground under foot, or suchlike, the activity of the fire immediately stops.

33. Being brought close to a warm body increases heat according to the degree of closeness; this also happens in the case of light; the closer an object is brought to light the more visible it is.

34. A combination of different heats increases heat unless there is a mixture of substances; for a big fire and a small fire in the same place increase each other's heat to some degree; but warm water poured into boiling water cools.

35. The duration of a hot body increases heat. For heat is constantly coming out and passing over and mixing with the preexisting heat, so that it multiplies the heat. A fire does not warm a room as much in half an hour as it does in the course of a whole day. This is not the case with light, for a lamp or a candle in a given spot gives no more light after a long time than it did right at the beginning.

36. Irritation by ambient cold increases heat; as one may see in the case of fires in bitter cold. We think that this occurs not so much because the heat is confined and contracted (which is a kind of union), but because it is exasperated, as when air or a cane is violently compressed or bent, it does not spring back to its former point but goes beyond it to the other side.

Make a careful experiment with a cane or something like that; put it into a flame, and see whether it is not burned more rapidly at the edge of the flame than in the centre.

37. There are several degrees of susceptibility to heat. First of all, note how even a little, weak heat alters and slightly warms things that are the least susceptible to heat. Even the heat of the hand gives some warmth to a small ball of lead or any metal held for just a short time. So easily is heat transmitted and aroused, and it happens in all substances without apparent change to any of them.

38. The readiest of all substances in our experience to take up and to lose heat is air. This is best seen in thermometers.[19] You make them as follows. Take a glass bottle with a rounded belly and a narrow, elongated neck; turn such a bottle upside down and insert it, mouth down, belly up, in another glass vessel which contains water, allowing the bottom of the receiving vessel to just touch the rim of the inserted bottle, and let the neck of the inserted bottle lie on the mouth of the receiving vessel and be supported by it; to do this more easily, place a little wax on the mouth of the receiving vessel; but do not completely seal up the mouth, lest a lack of incoming air should impede the movement we are to speak of; for it is a very light and delicate movement.

The upturned bottle, before being inserted in the other, should be warmed at a flame from above, i.e. on its belly. After the bottle has been placed there, as we said, the air (which had expanded because it was warmed) will withdraw and contract after the time it takes for the applied heat to be lost, to the extension or dimension which the ambient or outside air had at the time when the bottle was inserted, and will draw the water up to this extent. A long, narrow paper should be attached, marked with degrees (as many as you like). You will also observe, as the temperature of the day rises or falls, that the air contracts into a smaller space because of cold and expands into a wider area because of heat. This will be shown by the water rising when the air contracts and going down, or being forced down, when the air expands. The sensitivity of the air to cold and heat is so subtle and sensitive that it far surpasses the sensitivity of human touch; and a ray of sun or the warmth of a breath, to say nothing of the heat of a hand, placed over the top of the bottle, instantly sends the water down in a noticeable manner. We believe that animal spirit has a still more exquisite

[19] *vitrum calendare*

sense of heat and cold, but is dulled and obstructed by the mass of the body.

39. Most sensitive to heat after air, we think, are bodies which have been newly altered and compressed by cold, like snow and ice; for they begin to melt and thaw with just a gentle heat. After them, perhaps, quicksilver. Then come fatty substances, like oil, butter and so on; then wood; then water; finally stones and metals which are not easily warmed, especially internally. Once they have taken heat, however, they do retain it for a very long time; so that brick, stone or iron, once fired up and plunged and submerged in a basin of cold water, retains so much heat that it cannot be touched for (more or less) a quarter of an hour.

40. The less the mass of a body, the more quickly it warms up when placed next to a warm body; this shows that all heat in our experience is somehow opposed to tangible body.

41. To the senses and to human touch heat is a variable and relative thing; so that tepid water feels hot to a hand in the grip of cold, but if the hand warms up, it feels cold.

XIV

Anyone may easily see how poor our history is, since we are often compelled to make use in the above tables of the words 'do an experiment', or 'investigate further'; to say nothing of the fact that in place of proven history and reliable instances we insert traditions and tales (though not without noting their dubious authenticity and authority).

XV

We have chosen to call the task and function of these three tables the *Presentation of instances to the intellect*. After the *presentation* has been made, *induction* itself has to be put to work. For in addition to the *presentation* of each and every instance, we have to discover which nature appears constantly with a given nature or not, which grows with it or decreases with it; and which is a limitation (as we said above) of a more general nature. If the mind attempts to do this affirmatively from the beginning[20] (as it always does if left to itself), fancies will arise and conjectures and poorly defined

[20] Cf. 1.46, 105.

notions and axioms needing daily correction, unless one chooses (in the manner of the Schoolmen) to defend the indefensible. And they will doubtless be better or worse according to the ability and strength of the intellect at work. And yet it belongs to God alone (the creator and artificer of forms), or perhaps to angels and intelligences, to have direct knowledge of forms by affirmation, and from the outset of their thought. It is certainly beyond man, who may proceed at first only through *negatives* and, after making every kind of exclusion, may arrive at affirmatives only at the end.

XVI

Therefore we must make a complete analysis and separation of a nature, not by fire but with the mind, which is a kind of divine fire. The first task of true *induction* is the *rejection* or *exclusion* of singular natures which are not found in an instance in which the given nature is present; or which are found in an instance where the given nature is missing; or are found to increase in an instance where the given nature decreases; or to decrease when the given nature increases. Only when the *rejection* and *exclusion* has been performed in proper fashion will there remain (at the bottom of the flask, so to speak) an affirmative form, solid, true and well-defined (the volatile opinions having now vanished into smoke). This takes no time to say, but there are many twists and turns before one gets there. But we will hopefully leave out nothing that leads to this end.

XVII

When we seem to assign such an important role to forms, we must carefully caution and constantly warn, in case what we say is wrongly taken as referring to the kind of forms which have hitherto been familiar to men's thoughts and contemplations.[21]

First, we are not speaking at present of composite forms, which are (as we said) conjunctions of simple natures, in the common way of things, like lion, eagle, rose, gold and so on. It will be appropriate to deal with them when we get to *latent processes* and *latent structures*, and the uncovering of them as they are found in substances (so-called) or compound natures.

[21] Cf. 1.51, 65.

Again, what we have said should not be understood (even as far as simple natures are concerned) of abstract forms and ideas, which are not defined in matter or poorly defined. When we speak of forms, we mean simply those laws and limitations of pure act which organise and constitute a simple nature, like heat, light or weight, in every kind of susceptible material and subject. The form of heat therefore or the form of light is the same thing as the law of heat or the law of light, and we never abstract or withdraw from things themselves and the operative side. And so when we say (for example) in the inquiry into the form of heat, *Reject* rarity, or, rarity *is not of the form of* heat, it is the same as if we said, *Man can* superinduce *heat on a dense body*, or on the other hand, *Man can take away heat or bar it from a rare body*.

Our forms too may seem to someone to have something abstract about them, because they mix and combine heterogeneous elements (for the heat of heavenly bodies and the heat of fire seem to be very heterogeneous; the redness in a rose or such like is very different from that appearing in a rainbow or in the rays of an opal or diamond; and so are death by drowning, by burning, by a sword thrust, by a stroke and by starvation, and yet they are similar in having the nature of heat, redness and death). Anyone who thinks so should realise that his mind is captive and in thrall to habit, the surface appearance of things, and to opinions. For it is quite certain that however heterogeneous and foreign, they are similar in the form or law which defines heat or redness or death; and human power cannot be freed and liberated from the common course of nature, and opened up and raised to new effectiveness and new ways of operating, except by the uncovering and discovery of such forms. After this union of nature, which is absolutely the principal thing, we will speak later, in their place, of the divisions and veins of nature, both the ordinary divisions and those which are internal and more true.

XVIII

And now we must give an example of the *exclusion* or *rejection* of natures which are found by the *tables of presentation* not to belong to the form of heat; remarking in passing that not only are individual *tables* sufficient to *reject* a nature, but so is every one of the individual instances contained in them. For it is obvious from what I have said that every *contradictory instance* destroys a conjecture about a form. We do sometimes however provide two or three instances of an exclusion, for the sake of clarity, and to show more fully how the tables are to be used.

Example of *exclusion* or rejection *of natures* from the form of heat:

1. By the rays of the sun, *reject* elemental nature.

2. By ordinary fire, and particularly underground fires (which are furthest away and least affected by rays from the heavens), *reject* heavenly nature.

3. By the fact that bodies of all kinds (i.e. of minerals, vegetables, the external parts of animals, of water, oil, air and so on) are warmed by simply getting near to a fire or other warm body, reject variation or more or less subtle textures in bodies.

4. By heated iron and metals, which warm other bodies but are no way diminished in weight or substance, *reject* the attachment or admixture of the substance of another hot body.

5. By boiling water and air, and also by metals and other solids which have been warmed but not to the point of catching fire or redness, *reject* light and brightness.

6. By the rays of the moon and other stars (except the sun), again *reject* light and brightness.

7. By *comparison* with burning iron and the flame of spirit of wine (of which heated iron has more heat and less light, the flame of spirit of wine has more light and less heat), once again *reject* brightness and light.

8. By heated gold and other metals, which have the densest mass in the whole, *reject* rarity.

9. By air, which remains rare, however cold it gets, once more *reject* rarity.

10. By heated iron, which does not increase in size, but keeps the same visible dimension, *reject* local or expansive movement in the whole.

11. By the swelling of air in thermometers and the like, which moves in space and obviously in an expanding manner, and yet does not acquire obvious increase in heat, once again *reject* local movement or expansive movement in the whole.

12. By the easy warming of all bodies without any destruction or noticeable alteration, *reject* destructive nature or the violent addition of any new nature.

13. By the agreement and conformity of similar effects displayed by both heat and cold, *reject* both expanding and contracting motion in the whole.

14. By the generation of heat from rubbing bodies together, *reject* fundamental nature. By fundamental nature we mean one which is found existing in a nature and is not caused by a preceding nature.

There are also other natures; we are not composing complete tables, but only examples.

Not a single one of the natures listed comes from the form of heat. One need not be concerned with any of the natures listed in an operation on heat.

XIX

True *induction* is founded on *exclusion*, but is not completed until it reaches an affirmation. In fact an *exclusion* itself is not in any way complete, and cannot be so at the beginning. For *exclusion*, quite obviously, is the *rejection* of simple natures. But if we do not yet have good, true notions of simple natures, how can an *exclusion* be justified? Some of the notions we have mentioned are vague and poorly defined (e.g. the notion of an elementary nature, the notion of a heavenly nature, the notion of rarity). We recognise and always keep in mind how large a task we are undertaking (to make the human intellect equal to things and to nature) and therefore do not stop at the present stage of our teaching. We go further, and devise and provide more powerful aids for the use of the intellect; and these we now set out. In *the interpretation of nature*, surely, the mind has to be formed and prepared to be content with an appropriate degree of certainty, and yet to recognise (especially at the beginning) that what is before us depends heavily on what is to come.

XX

And yet because truth emerges more quickly from error than from confusion, we believe it is useful to give the intellect permission, after it has compiled and considered three tables of *first presentation* (as we have done), to get ready to try an *interpretation of nature* in the affirmative on the basis of the instances in the tables and of instances occurring elsewhere. We have chosen to call such a first attempt an *authorisation of the intellect*, or a *first approach to an interpretation*, or a *first harvest*.

A first harvest of the form of heat

Notice (as is quite clear from what I have said) that the form of a thing is in each and every one of the instances in which the thing itself is; otherwise it would not be a form: and therefore there can be absolutely no contradictory *instance*. And yet the form is much more obvious and evident in some instances than in others, namely in the instances where the nature of the form is less checked, obstructed and limited by other natures. We have chosen to call such instances *conspicuous* or *revealing instances*.[22] Let us proceed then to the actual *first harvest* of the form of heat.

In each and every instance, the nature of which heat is a limitation seems to be motion. This is most apparent in a flame, which is always in motion; and in boiling or bubbling liquids, which are also always in motion. It appears also in the intensifying or increase of heat produced by motion; as by bellows and winds (see Instances 29 of Table 3). And similarly with motion of other kinds (see Instance 28 and 31 of Table 3). It is apparent once more in the extinction of fire and heat by any power- ful compression which puts the brake on motion and makes it cease (see Instances 30 and 32 of Table 3). It is also apparent in the fact that every body is destroyed or at least significantly altered by any fire or strong and powerful heat; hence it is quite obvious that in the internal parts of a body, heat causes tumult, agitation and fierce motion which gradually brings it to dissolution.

What we have said about motion (i.e. that it is like a *genus* in relation to heat) should not be taken to mean that heat generates motion or that motion generates heat (though both are true in some cases), but that actual heat itself, or the quiddity of heat, is motion and nothing else; limited how- ever by the *differences* which we shall lay out shortly, after adding some caveats to avoid ambiguity.

Heat as felt is a relative thing, and is not universal but relative to each individual; and it is rightly regarded as merely the effect of heat on the animal spirit. Further it is in itself a variable thing, since the same object gives rise to a perception of both heat and cold (according to the condition of the senses), as is clear from Instance 41 of Table 3.

The form of heat should not be confused with the communication of

---

[22] See 11.24.

heat or its conductive nature, by which one body is warmed by contact with another body which is hot. Heat is different from warming. Heat is produced by a movement of rubbing without any preceding heat, and this excludes warming from the form of heat. Even when heat is produced by closeness to heat, the effect is not due to the form of heat but depends wholly on a higher and more common nature, viz. on the nature of assimilation or multiplication, which requires a separate investigation.

Fire is a popular notion without value: it is made up of a union of heat and light in a body, as in an ordinary flame, and in bodies heated till they are red.

Having removed all ambiguity, we must now come at last to the true *differences* which limit motion and constitute it as the form of heat.

The *first difference* is that heat is an expansive motion, by which a body seeks to dilate and move into a larger sphere or dimension than it had previously occupied. This difference is most obvious in a flame; here the smoke or cloudy exhalation visibly broadens and opens out into a flame.

It is also apparent in all boiling liquids, which visibly swell and rise and give off bubbles; and it pursues the process of its own expansion until it turns into a body which is far more extensive and broadened than the liquid itself, i.e. into steam or smoke or air.

It is also apparent in wood and every kind of combustible thing, where there is sometimes sweating and always evaporation.

It is also apparent in the melting of metals, which (being of highly compact substance) do not easily swell and dilate; their spirit first dilates in itself and so conceives a desire for still greater dilation, then visibly thrusts and forces the more solid parts into liquid form. If the heat is still further intensified, it dissolves and turns much of it into a volatile substance.

It is apparent also in iron or rocks; although they do not melt, they are not fused, but they are softened. This also happens with sticks of wood; they become flexible when gently warmed in hot ashes.

But this motion is best seen in air, which under the influence of a little heat dilates itself immediately and perceptibly; as by Instance 38 of Table 3.

It is also apparent in the contrary nature, that of cold. For cold contracts every substance and forcibly narrows it; so that in spells of intense cold nails fall out of walls, bronze objects split, and glass which has been warmed and then suddenly plunged into cold cracks and breaks. Air similarly withdraws into a smaller space under the influence of a little cooling; as by Instance 38 of Table 3. But we will speak more fully of this in the inquiry on cold.

And it is no wonder if heat and cold exhibit several similar actions (on which see Instance 32 of Table 2), since two of the following *differences* (of which we are about to speak) belong to both natures; though in this *difference* (of which we are now speaking) the actions are diametrically opposite. For heat gives an expansive, dilating motion, and cold gives a contracting, shrinking movement.

The *second difference* is a variation of the first; namely that heat is an expansive motion, or motion towards a circumference, under the condition that the body rises with it. For there is no doubt that there are many mixed motions. For example, an arrow or a javelin rotates as it flies and flies in rotating. Similarly too the motion of heat is both an expansion and a movement upwards.

This *difference* is apparent in a pair of tongs or an iron poker put in the fire: because if you put it in upright holding it with your hand from above, it quickly burns your hand; but if you put it in from the side or from below, it is much slower to burn the hand.

It is conspicuous also in distillation by means of a retort, which is used for delicate flowers which easily lose their scent. Trial and error has found that one should place the flame above rather than beneath, so as to burn less. For all heat, and not just a flame, rises.

On this subject try an experiment with the contrary nature of cold, as to whether cold does not contract a body by descending downwards just as heat dilates a body by ascending upwards. Take two iron rods or two glass tubes (in other respects equal) and heat them somewhat; place a sponge full of cold water or snow under one and over the other. Our view is that the cooling at its ends will be quicker in the rod with the snow above it than in that with the snow beneath it; contrary to what happens in the case of heat.

The *third difference* is that heat is a motion which is not uniformly expansive throughout the whole of a body, but expansive through its smaller particles, and is at once checked and repelled and bounced about, so that it takes on a back-and-forth motion, always scurrying about, and straining and struggling, angry at the beating it takes; hence the fury of fire and heat.

This *difference* is most apparent in a flame and in boiling liquids; which are incessantly agitated, swelling in small spots and subsiding again.

It is also apparent in bodies which are so tough and compact that they do not swell or gain in mass when they are heated or fired; like heated iron, in which the heat is very fierce.

It is apparent also in a hearth fire burning brightest in the coldest weather.

It is also apparent in the fact that no heat is observable when air expands in a calendar glass without obstacle or counter-pressure, i.e. uniformly and equally. And no particular heat is observable in the case of winds which have been shut up and then burst forth with great violence; that is because the motion is of the whole without a back-and-forth motion in the parts. Try an experiment on this, whether a flame does not burn more fiercely towards the edges than in the middle.

It is apparent also in the fact that all burning passes through the minute pores of the body on fire; so that the burning pits, penetrates, stabs and pricks like a thousand needle points. This is also the reason why all strong waters (if akin to the body in which they act) have the effect of fire because of their corrosive, piercing nature.

The *difference* of which we are now speaking is shared with the nature of cold: in cold the contractive motion is checked by the contrary pressure to expand, while in heat the expansive motion is checked by the contrary pressure to contract.

Therefore whether they penetrate the parts of the body towards the interior or towards the exterior, the explanation is the same, though their strength is quite different in the two cases. For we do not experience on the surface of the earth anything exceedingly cold. See Instance 27 of Table 1.

The *fourth difference* is a variation of the previous one. It is that the motion of pricking and penetration must be quite rapid, not slow, and takes place at the level of the particles, minute as they are; and yet not the very smallest particles, but those which are somewhat larger.

This difference is apparent from a comparison of the effects which fire gives with those made by time or age. For age or time withers, consumes, subverts and turns to dust no less than fire; or rather much more subtly. But because such motion is very slow and by very tiny particles, no heat is observable.

It is apparent also in the comparison between the dissolution of iron and of gold. Gold is dissolved without arousing heat, but iron with a violent arousal of heat, though in a quite similar length of time. The reason is that in gold the entrance of the separating liquid is gentle and discreet, and the particles of gold yield easily; but in iron the entry is rough and forcible and the particles of iron are more stubborn.

It is apparent also to some extent in some gangrenes and flesh rotting, which cause little heat or pain because the putrefaction is delicate.

And this is the *first harvest* or *preliminary interpretation* of the form of heat, made by *the leave given to the intellect*.

On the basis of this *first harvest*, the true form or definition of heat (of heat as a universal notion, not relative merely to sense) is, in a nutshell, as follows: *heat is an expansive motion which is checked and struggling through the particles.* And expansion is qualified: *while expanding in all directions it has some tendency to rise.* And the struggle through the particles is thus qualified: *it is not completely sluggish, but excited and with some force.*

The thing is the same as far as operation is concerned. This is the sum of it: *If in any natural body you can arouse a motion to dilate or expand; and if you can check that motion and turn it back on itself so that the dilation does not proceed equally but partly succeeds and is partly checked, you will certainly generate heat.* It is irrelevant whether the body is elementary (so-called) or imbued with heavenly substances; whether luminous or opaque; whether rare or dense; whether spatially expanded or contained within the bounds of its first size; whether tending toward dissolution or in a steady state; whether animal, vegetable or mineral, or water, oil or air, or any other substance whatsoever which is capable of the motion described. Heat to the senses is the same thing; but with the analogy that belongs to our senses. But now we must proceed to other aids.

XXI

After the *Tables of first presentation*, after *rejection or exclusion*, and after making the *first harvest* on the basis of them, we must proceed to the other aids to the intellect in the *interpretation of nature* and in true and complete *induction*. In setting them out we shall continue to use heat and cold when we need tables, but where we want just a few examples, we shall make use of any other examples, so that we may give a wider scope to our teaching without confusing the inquiry.

We shall speak then in the first place of *privileged instances*;[23] secondly of *supports for induction*; third of *the refinement of induction*; fourth of the adaptation of the *investigation to the nature of the subject*; fifth of *natures which are privileged* so far as investigation is concerned, or of which inquiries we make first and which ones later; sixth of the *limits of investigation*, or of a summary of all natures universally; seventh of *deduction to practice*, or of how it relates to man; eighth of *preparations for investigation*; and finally of the *ascending and descending scale of axioms*.

XXII

Among *privileged instances* we shall first bring forward *solitary instances*.[24] *Solitary instances* are those which exhibit the nature under investigation in subjects which have nothing in common with other subjects but that very nature; or again which do not exhibit the nature under investigation, in subjects which are similar in all things to other subjects except in that nature. It is evident that instances of this kind cut out the rambling, and are a quick route to confirming an *exclusion*, so that a few of them are as good as many.

For example: in an investigation of the nature of *colour*, prisms and crystals, and also dew and such things, which make colours in themselves and throw them outside themselves onto a wall, are *solitary instances*. For they have nothing in common with the inherent colours in flowers, coloured gems, metals, woods etc., except colour itself. Hence it is easy to infer that colour is nothing other than the modification of a ray of light admitted and received, in the first case through different degrees of incidence, in the second through various textures and structures of body. And these are *solitary instances* concerned with resemblance.

Again, in the same inquiry, distinct veins of black and white in marble, and variations of colour in flowers of the same species, are *solitary instances*.

[23] *praerogativae instantiarum*: so named from the *centuria praerogativa*, the aristocratic section of the *comitia centuriata* at Rome, which had the privilege of voting first at meetings and of announcing its vote before the other 'centuries' voted; it thus indicated to the others which way to vote.
[24] *instantiae solitariae*

For the white and the black in marble and the patches of red and white in carnations agree in almost everything except the colour itself. Hence the easy conclusion that colour does not have much in common with the intrinsic natures of a body but lies only in the grosser, quasi-mechanical positioning of the parts. And these are *solitary instances* concerned with difference. We have chosen to call both kinds *solitary instances*; or *wild ones*,[25] a term taken from the astronomers.

XXIII

The *privileged instances* which we will put second are *instances of transition*.[26] These are the instances in which the nature which we are looking for, if previously non-existent, is in transition to being, and if already existing, towards non-being. And therefore in both these opposite movements such instances are always double; or rather it is one instance prolonged in its motion or passage to the opposite point of the circle. Such instances are not only a quick way to confirm an *exclusion*, but also pin down the *affirmation*, or actual *form*, to a small area. For the form has to be something which is introduced by one kind of transition or on the other hand removed and annihilated by another kind. And though every exclusion encourages an *affirmation*, this is still achieved more directly in the same subject than in different subjects. And the form (as is abundantly clear from my discussion) which reveals itself in the one case leads us on to all of them. The simpler the transition, the more highly we should value the instance. Again *instances of transition* are quite useful for the operative function; for when they exhibit the form combined with the cause that makes it so or prevents it from being so, they turn a bright light on an activity in some things, from which it is also an easy transition to the next thing. But there is a danger in them which requires a caution; they may bring the form too close to the efficient cause, and may soak, or at least dip, the intellect in a false view of the form in relation to the efficient cause. The efficient cause is always defined as nothing other than the vehicle or bearer of the form. There is an easy remedy to this problem in a properly conducted exclusion.

We should now give an example of an *instance of transition*. Let the nature to be sought be White or Whiteness; an instance in transition to produce

[25] Reading *ferales*, as Ellis proposes.
[26] *instantiae migrantes*

it is unbroken glass and powdered glass; also plain water and water stirred to a foam. For unbroken glass and plain water are transparent, not white; but powdered glass and foaming water are white, not transparent. And so one has to ask what happened to the glass or the water as a result of the transition. For it is obvious that the form of whiteness is imported and introduced by the pounding of the glass and the stirring of the water. Nothing else is found to have occurred but the fragmentation of the glass and the water into little bits and the introduction of air. And it was no small step towards the discovery of the form of whiteness that two bodies which are in themselves transparent more or less (namely air and water, or air and glass) exhibit whiteness as soon as they are made into minute fragments, because of the unequal refraction of rays of light.

But on this question one should also give an example of the danger and of the caution we mentioned. No doubt a mind which has been led astray by that sort of efficient cause will too easily conclude that air is always necessary to the form of whiteness, or that whiteness is only generated by transparent bodies; which are all completely false, and proven to be so by many exclusions. In fact it will rather be apparent (setting aside air and suchlike) that bodies which are quite even (in the parts of them that affect vision) yield transparency, while bodies which are unequal and of simple texture yield whiteness; bodies which are unequal and of compound but regular structure yield other colours except black; and bodies which are unequal and of compound but altogether irregular and disorderly structure yield black. And so an example has been given of an *instance of transition* towards the generation of whiteness in a nature under investigation. An *instance of transition* towards non-being in the same nature of whiteness is disintegrating foam or melted snow. Water sheds whiteness and takes on transparency after becoming whole without air.

Nor should we in any way fail to include among *instances of transition* not only instances moving towards generation and passing away but also those tending towards increase and diminution; since these too tend to reveal a form, as is abundantly clear from the definition of a form given above and the *table of degrees*. And thus there is an explanation analogous to that in the instances given above, why paper, which when dry is white, is less white when wetted (because of the exclusion of air and the reception of water) and tends more towards transparency.

XXIV

As third of the *privileged instances* we shall put *revealing instances*,[27] which we have mentioned already in the *first harvest on Heat*;[28] we also call them *conspicuous instances* or *liberated* or *dominant instances*. They are instances which reveal the nature under investigation naked and independent, and also at its height and in the supreme degree of its power; that is, liberated and free of impediments, or at least prevailing over them by the strength of its virtue, and suppressing and restraining them. Every body is susceptible of many forms of combination and compounding of natures they dull, depress, break and bind each other; and the individual forms are obscured. But some subjects are found in which the nature under investigation stands out from the others in its vigour, either because there are no obstacles or because its virtue is dominant. Such *instances* are particularly *revealing* of form. But even in these instances caution is necessary, and we have to restrain the intellect's hastiness. We should be suspicious of anything that obtrudes a form on us, and makes it leap out at us, so that it seems simply to spring to mind; we must insist on a strict and careful *exclusion*.

For example: suppose the nature is heat. The *revealing instance* of the motion of expansion, which (as I said above) is a thermometer of air. While a flame clearly shows expansion, it does not reveal the progress of expansion, because of its immediate extinction. Boiling water also does not reveal the expansion of water so well in its own body, because of the quick conversion of water into steam and air. And red-hot iron and suchlike are very far from revealing a progress; on the contrary, the suppression and fragmentation of the spirit by its dense and compacted particles (which tame and bridle the expansion) prevent the actual expansion from being fully apparent to the senses. But a thermometer clearly reveals expansion in air, and reveals it as conspicuous, progressive, enduring and not transient.

For another example, suppose the sought nature is weight. The *revealing instance* of weight is quicksilver. It far exceeds everything in weight except gold, which is slightly heavier. Quicksilver is a better instance than gold for revealing weight; because gold is solid and compact, which seems to be due to its density, whereas quicksilver is liquid and swelling with spirit, and yet far exceeds diamond in weight and things which are taken

[27] *instantiae ostensivae*
[28] At II.20.

to be the most solid. This reveals that the form of heaviness or weight is governed simply the amount of matter and not by how compact it is.

XXV

As fifth of the *privileged instances* we put *concealed instances*,[29] which we have also chosen to call *instances of the twilight*. They are almost the opposite to *revealing instances*. For they exhibit the nature under investigation at its lowest strength, as if in its origins or earliest efforts, tentative and trying itself out, but concealed beneath a contrary nature and subdued by it. Yet such instances are of the utmost importance for discovering forms, because like *revealing instances* they easily lead to differences, so concealed *instances* are the best guides to *genera*, that is, to those common natures of which the natures under investigation are simply the limitations.

For example: suppose the sought nature under investigation is solidity or the determinate, the opposite of which is the liquid or the fluid. *Concealed instances* are those which exhibit a very low and weak degree of solidity in a fluid, a bubble of water, for example, which is like a kind of solid and determinate skin made of the substance of water. Similarly with dripping water: if the water keeps on coming, it forms a thin thread so that the stream of water is unbroken; but if there is not enough water to keep on coming, it falls in round drops, which is the shape which best sustains the water against breakup. But at the actual moment when it ceases to be a thread of water and begins to fall in drops, the water itself leaps back up to avoid breaking. In metals which in the molten state are liquid but very viscous, the molten drops often leap back up and stay there. Much the same is true of children's mirrors, which small children make from spittle on reeds; here too one sees a solid film of water. But this is exhibited much better in the other children's game where they take water, make it a bit more viscous with soap and blow it up through a hollow straw, and thus make the water into something like a reservoir of bubbles; and through the admixture with air it takes on[30] such solidity that it may be tossed some distance into the air without breaking up. This is best seen in foam and snow, which take on such solidity that they can almost be cut; and yet both bodies are formed from air and water, which are both liquids. All of which quite clearly indicates that liquid and solid are mere vulgar notions adapted to the senses; that in reality there is in all

[29] *instantiae clandestinae*
[30] Reading *induit* for *inducit*.

bodies a tendency to avoid and evade breaking up; that this is weak and feeble in homogeneous substances (which liquids are) but more lively and powerful in bodies compounded of heterogeneous substances; the reason is that the addition of heterogeneity unites bodies, while the entrance of the homogeneous dissolves and disbands them.

Another example: suppose the nature under investigation is Attraction, or the or the coming together of bodies. The most remarkable *revealing instance* of its form is the magnet. The nature contrary to the attracting nature is the non-attracting nature, even in the same substance. Iron, for example, which does not attract iron, just as lead does not attract lead, nor wood wood, nor water water. A *concealed secret* is a magnet armed with iron, or rather iron in an armed magnet. Its nature is such that an armed magnet does not attract iron at a distance more powerfully than an unarmed magnet. But if the iron is brought near enough to touch the iron in the armed magnet, then the armed magnet holds a much greater weight of iron than a simple unarmed magnet, because of the similarity of substance, iron against iron; this activity was completely *concealed* and latent in the iron before the magnet approached. And thus it is clear that the form of coming together is something which is lively and strong in the magnet, weak and latent in the iron. Likewise it has been remarked that small wooden arrows without iron tips shot from large carbines penetrate wooden objects (e.g. the sides of ships, or suchlike) more deeply than the same arrows tipped with iron, because of the similarity of the substances (wood on wood), though this was previously hidden in the wood. Likewise, though whole bodies of air do not obviously attract air or water water, yet a bubble which touches another bubble more easily dissolves it than if there were no second bubble, because of the inclination that water has to come together with water, and air with air. Such *Concealed Instances* (which are remarkably useful, as I have said) give the best glimpse of themselves in small portions of substances. For larger masses of things follow more universal and general forms, as will be explained in due course.

XXVI

As fifth of the *privileged instances* we shall put *constitutive instances*;[31] which we have also chosen to call *bundled instances.* These are instances which

---

[31] *instantiae constitutivae*

constitute one species of a nature under investigation as a Lesser Form. Genuine forms (which are always convertible with the natures under investigation) are hidden in the depths and not easily discovered, and therefore the thing itself and the feebleness of human understanding require that we should not neglect but carefully observe particular forms which group certain bundles of instances (though by no means all) together into a common notion. For whatever unites a nature, even imperfectly, opens a way to the discovery of forms. And therefore instances which are useful for this purpose have considerable value but no privilege.

However, one must use the greatest caution here lest the human intellect, after finding a few of these particular forms and making partitions or divisions of the nature under investigation, rest in them altogether and not prepare itself for the true discovery of the great Form, but assume that the nature is radically multiple and divided, and spurn and reject any further unity in the nature as a thing of superfluous subtlety which verges on the merely abstract.

For example: suppose the nature under investigation is memory, or that which provokes and assists memory. The *constitutive instances* are order or arrangement, which clearly helps memory; and also in artificial memory the constitutive instances are 'places'.[32] 'Places' may be either places in the literal sense, such as a door, a corner, a window and so on, or familiar and well-known persons, or may be anything at all (provided they are put in a certain order), such as animals or herbs; also words, letters, characters, historical persons etc., though some of these are more suitable than others. Such 'places' give remarkable assistance to the memory, and lift it well above its natural powers. Likewise poetry stays more easily in the mind and is more easily learned than prose. And one species of assistance to the memory is constituted from this *bundle* of three instances, i.e. order, the artificial memory of 'places' and verses. This species may rightly be called *curtailment of the unlimited*. For when one attempts to recall something or bring it to mind, if he has no prior notion or conception of what he is looking for, he is surely looking, struggling and running about here and there in a seemingly *unlimited* space. But if he has a definite notion, the ulimited is immediately curtailed and the range of the memory is kept within bounds. And there is a clear and definite notion in the three instances given above. In the first there should be something which agrees

---

32 Loci = the Greek, *topoi*, as in the 'commonplaces' of rhetoric.

with order; in the second there should be an image which has some relation or congruence with those specific 'places'; in the third, there should be words which have the rhythm of the verse. Thus the unlimited is curtailed. Other instances will yield another species: anything that makes an intellectual notion strike the senses assists the memory (this is the most prevalent method in artificial memory). Other instances will yield another species: the memory is assisted by anything that makes an impression on a powerful passion, inspiring fear, for example, or wonder shame or joy. Other instances will yield another species: things which are imprinted on[33] a mind which is clear and uncluttered either before or after, for example what we learn in childhood or what we think of before we go to sleep, or the first experience of a thing, are more likely to remain in the memory. Other instances yield this next species: a large variety of circumstances or devices help the memory, e.g. breaking a text up into sections, or reading or reciting aloud. Other instances, lastly, will yield this final species: things which are expected and attract attention stay better than those which just slip by. Hence if you read a piece of text through twenty times, you will not learn it by heart so easily as if you read it ten times while attempting to recite it from time to time and consulting the text when your memory fails. Thus there are about six Lesser Forms of things which help the memory: namely curtailment of the unlimited; reduction of the intellectual to the sensual; impression on a strong passion; impression on a clear mind; a large variety of devices; and anticipation.

Another similar example: suppose the nature under investigation is taste or a tasting. The *instances* which follow are *constitutive*: namely, people who cannot smell and are bereft of that sense by nature fail to notice or distinguish by taste rancid or rotten food, or on the other hand food cooked in garlic or rosewater, or suchlike. People whose nostrils are accidentally blocked by phlegm running down through them fail to notice or distinguish anything which is off or rancid or sprinkled with rosewater. But if they are afflicted with such phlegm and give their nose a good blow at the actual moment when they have the rancid or perfumed thing in their mouths or on their palate, they have at that moment a clear perception of the rancidity or perfume. These instances will yield and constitute this species, or rather this part, of taste; so that the sensation of tasting is in part simply an interior smelling, which passes down from the higher nasal tubes

---

[33] Reading *in* for *a*.

to the mouth and palate. On the other hand, all salty tastes, and sweet, sharp, acid, sour and bitter tastes, and so on, give the same sensation to one whose sense of smell is absent or blocked as to anyone else. Hence it is clear that the sense of taste is a composite of interior smell and a kind of exquisitely sensitive touch; but this is not the place to discuss that.

Another similar example: suppose the nature under investigation is the passage of a quality without mixture of substance. The instance of light will yield or constitute one kind of passage; heat and magnetism another. For the passage of light is virtually instantaneous, and dies directly the source light is removed. But heat and magnetic power are transmitted or rather aroused in another body, and then stay and persists for quite a long time after the source is removed.

Finally, the privilege of *constitutive instances* is very great indeed, in that they contribute most to forming both definitions (especially particular definitions) and divisions or partitions of natures. Plato put it rather well: *the man who knows well how to define and divide should be regarded as a god.*

XXVII

As the sixth of the privileged instances we will put *instances of resemblance*[34] or *analogous instances*, which we have also chosen to call *parallels*, or *physical similarities*. They are instances which reveal similarities and connections between things, not in the lesser forms (which is the role of *constitutive instances*), but in the actual, concrete object. Thus they are like the first and lowest steps towards the unity of a nature. They do not establish any axiom directly from the beginning, but only indicate and point to some agreement between bodies. But though they do not contribute much towards the discovery of forms, they are extremely useful in uncovering the structure of parts of a whole, and perform a kind of dissection upon its members; accordingly they sometimes bring us step by step to sublime and noble axioms, particularly axioms relating to the structure of the world rather than to simple forms and natures.

Some examples of *instances of resemblance* are the eye and a mirror, the structure of the ear and echoing places. Apart from the actual observation of the likeness, which has several uses in itself, it is a simple matter to produce and form the following axiom from this resemblance: that the

---

[34] *instantiae conformes*

organs of sense are of a similar nature to bodies which give off reflections to the senses. Taking its cue from this fact, the understanding in its turn rises without difficulty to a loftier and nobler axiom: namely that agreements or sympathies between bodies endowed with sense and between inanimate objects without sense differ only in the fact that in the former case an animal spirit is present in a body equipped to receive it, but is lacking in the latter. Consequently there might be as many senses in animals as there are agreements in inanimate bodies, if there were perforations in the animate body for the diffusion of animal spirit into a duly prepared limb, as into a suitable organ. And there undoubtedly are as many motions in an inanimate body without animal spirit, as there are senses in animals, though there have to be more motions in inanimate bodies than senses in animate bodies, because there are so few organs of sense. An obvious and ready example of this is pain. Though there are many kinds of pain in animals, with different characteristics (the pains of burning, of intense cold, of pricking, pressing, stretching and so on, are quite different from one another), it is quite certain that, so far as they are motion, they all occur in inanimate bodies; as in wood or rock when it is burned or contracted by cold, or bored, cut, bent or crushed; and so for other things; though the senses do not come into it because of the absence of animal spirits.

Likewise the roots and branches of plants (odd as it may be to say so) are instances of resemblance. For everything which is vegetable swells and puts out limbs into its environment both upwards and downwards. The only difference between roots and branches is that the root is shut in the earth and the branches are exposed to air and sun. Take a tender, living shoot of a tree and bend it over and stick it into a clump of earth, even if it is not attached to the ground itself, and it will immediately produce a root not a branch. If on the other hand a plant is covered with earth but so obstructed by a rock or hard substance that it is prevented from developing foliage on top, it will put out branches in the air below.

Tree resins and most rock gems are also *instances of resemblance*. For both are simply exudations and percolations of juices: in the first case of juices from trees, in the second from rocks; so that both are made clear and lustrous by close, careful percolation. This is also the reason why the hide of animals is not so pretty and highly coloured as the plumage of many birds, because juices do not percolate so delicately through skin as they do through quills.

Other *instances of resemblance* are the scrotum in males and the womb in females. Hence that remarkable structure which differentiates the sexes (so far as land animals are concerned) appears to be a matter of outside and inside; because the greater force of heat in the masculine sex forces the genitals outside, whereas in females the heat is too weak to do this, with the result that they remain inside.

Likewise the fins of fish and the feet of quadrupeds or the feet and wings of birds, are *resembling instances*; and Aristotle added the four undulations in the motion of snakes. And thus in the general structure of things the motion of living creatures often seems to depend on sets of four joints or bends.

Likewise teeth in land animals and birds' beaks are *instances of resemblance*; from which it is clear that in all finished animals a kind of hard substance gathers at the mouth.

Likewise it is not absurd that there should be similarity and resemblance between a man and an inverted plant. For the head is the root of the nerves and faculties of animals; and the seed parts are at the bottom (ignoring the extremities of legs and arms), whereas in a plant the root (which is like the head) is regularly located at the bottom, and the seeds are at the top.

Finally, we much absolutely insist and often recall that men's attention in the research and compilation of natural history has to be completely different from now on, and transformed to the opposite of the current practice. Up to now men have put a great deal of hard and careful work into noting the variety of things and minutely explaining the distinctive features of animals, herbs and fossils; most of which are more sports of nature than real differences of any use to the sciences. Such things certainly give pleasure and even sometimes have practical use, but contribute little or nothing to an intimate view of nature. And so we must turn all our attention to seeking and noting the resemblances and analogies of things, both in wholes and in parts. For those are the things which unite nature, and begin to constitute sciences.

But in all this one has to be strict, and very cautious, and accept as resembling and *analogous* only those *instances* which denote physical similarities (as we have said from the beginning); that is, real and substantial similarities which are founded in nature, not accidental and apparent similarities; much less the sort of superstitious or curious resemblances constantly featured by writers on natural magic (light-minded men and hardly worth mentioning in such serious matters as we are now discussing) when with much vanity and foolishness they describe, and even sometimes invent, empty similarities and sympathies in things.

These things aside, *instances of resemblance* should not be ignored in bigger matters, even in the actual configuration of the earth; such as Africa and the Peruvian region with the coastline extending to the Strait of Magellan. For both regions have similar isthmuses and similar promontories, and that does not happen without a reason.

Likewise the New and the Old World: in the fact that both worlds are immensely wide towards the North and narrow and pointed towards the South.

Quite remarkable *instances of resemblance* are the intense cold in what they call the middle region of the air, and the very fierce fires which are often seen to erupt from places under the earth; two things which are ultimates and extremes: the extremes of the nature of cold towards the circuit of the sky and of the nature of heat towards the bowels of the earth; by opposition, or rejection of the contrary nature.

Finally, *resemblance of instances* in the axioms of science deserves notice. The rhetorical trope called *contrary to expectation* resembles the musical figure known as avoidance of the cadence. Similarly, the mathematical postulate that 'things which are equal to a third thing are also equal to each other', corresponds with the structure of the syllogism in logic, which joins things which agree in a middle term. Finally, it is very useful that as many people as possible should have a keen sense for tracing and tracking physical similarities and resemblances.

XXVIII

Seventh among the privileged instances we shall put *unique instances*;[35] which we have also chosen to call *irregular* or *heteroclite* instances (borrowing a term from the grammarians). These are instances which reveal in concrete form bodies which seem to be extraordinary and isolated in nature, having very little in common with other things of the same kind. For *instances of resemblance* are like one another, but *unique instances* are *sui generis*. The use of unique instances is like the use of *concealed instances*, i.e. to raise and unite nature for the purpose of discovering kinds or common natures, which are afterwards to be limited by means of genuine *differentiae*. The inquiry should proceed until the properties and qualities found in things which can be regarded as wonders of nature are reduced

---

[35] *instantiae monodicae*

and comprehended under some specific form or law. Thus every irregularity or peculiarity will be found to depend on some common form; and the wonder at last lies merely in the minute *differentiae*, and in the degree and the unusual combination, not in the species itself; whereas now men's reasoning gets no further than to call these things secrets of nature or monstrosities, things without cause and exceptions to the general rules.

Examples of unique instances are the sun and the moon among the stars; the magnet among stones; quicksilver among metals; the elephant among quadrupeds; the sensation of sex among kinds of touch; keenness of scent in dogs among kinds of smell. Also the letter S is taken as unique by grammarians, because of the easy way in which it combines with consonants, sometimes two, sometimes three; as no other letter does. Such instances should be prized, because they sharpen and quicken inquiry, refreshing a mind staled by habit and by the usual course of things.

XXIX

Eighth among *privileged instances* we shall put *deviant instances*,[36] that is errors of nature, freaks and monsters, where nature deflects and declines from its usual course. Errors of nature differ from *unique instances* in the fact that unique instances are wonders of species, whereas errors of nature are wonders of individuals. But their use is pretty much the same, because they fortify the intellect in the face of the commonplace, and reveal common forms. Here too we must pursue the inquiry until the cause of the deviation is discovered. However, the cause does not succeed in becoming a form, merely a *latent process* on the way to a form. He who knows the ways of nature will also more easily recognise the deviations. And conversely he who recognises the deviations will more accurately describe the ways.

They also differ from unique instances in the fact that they much better prepare the way to practice and application. For it would be very difficult to create new species; but less difficult to vary known species and thus to develop many rare and unusual things. It is an easy transition from the wonders of nature to the wonders of art. For once a nature has been observed in its variation, and the reason for it made clear, it will be an easy matter to bring that nature by art to the point it reached by chance. And not just to that point but to other ends too; for errors in one direction

[36] *instantiae deviantes*

show and point the way to errors in and deviations in all directions. Here there is no need of examples, because there are so many. We must make a collection or particular natural history of all the monsters and prodigious products of nature, of every novelty, rarity or abnormality in nature. But this must be done with the greatest discretion, to maintain credibility. We shall especially suspect things that depend in any way on religion, like the prodigies in Livy; as well as what we find in writers on natural magic or alchemy or men of that sort who have a passion for fables. Facts should be taken from serious and credible history and from reliable reports.

XXX

Ninth *among privileged instance* we shall put *borderline instances*,[37] which we have also chosen to call *instances of sharing*.[38] They are instances which exhibit species of body which seem to be compounded of two species, or *elements*, that lie between one species and another. These instances may rightly be regarded as *unique* or *heteroclite instances*, since they are rare or extraordinary in the overall scheme of things. Nevertheless, they should be classed and discussed separately because of their value: they are excellent indicators of the composition and structure of things, they point to the reasons for the number and quality of regular species in the world, and they lead the intellect from what is to what can be regarded as *unique* or *heteroclite instances*, since, in the whole of nature, they are rare and extraordinary.

Examples are: moss, which is between putrefaction and plant; some comets, between stars and blazing meteors; flying fish, between birds and fish; bats, between birds and quadrupeds; also 'the ape, repulsive creature, how like us';[39] and mongrel animal offspring, as well as mixtures of different species; and the like.

XXXI

As tenth of the *privileged instances* we shall put *instances of power*,[40] or of the *sceptre* (to borrow a word from the emblems of government), which we have

---

[37] *instantiniae limitaneae*

[38] *Participia* continues the grammatical analogy found in the names for some of the other instances, e.g. 'irregular' or 'heteroclite' for the seventh set of instances. A *participium* is a participle, so called because participles 'share' the nature of both nouns and adjectives.

[39] Ennius, quoted by Cicero, *On the Nature of the Gods*, 1.35.

[40] *instantiae potestatis*

also chosen to call *man's contrivances* or *tools*. They are the noblest and most perfect works, the finished products of every art. For as the chief thing is that nature should contribute to human affairs and human advantage, the first step towards this end is to note and enumerate the works that are already within man's power (the provinces already occupied and subdued); particularly the most refined and finished works, because they provide the easiest and quickest route to new things not yet discovered. If one were to reflect carefully on them and then make a keen and persistent effort to develop the design, he would surely either improve it in some way, or modify it to something closely related, or even apply and transfer it to a still nobler purpose.

This is not the end of it. As rare and unusual works of nature arouse and stimulate the intellect to seek and discover forms capacious enough to contain them, so too do outstanding and admirable works of art; and more so, because the way of effecting and achieving such wonders of art is mostly quite plain, whereas in the wonders of nature it is often quite obscure. But here too we must be very careful not to hold down the mind and tether it to the ground.

For there is a danger that such works of art, which seem like the peaks and high points of human endeavour, may stun the intellect, bind it, and cast their own particular spell on it so that it becomes incapable of further knowledge, because it thinks that nothing of that kind can be done except in the same way as those things were done, with just a little more work and more careful preparation.

To the contrary, it must be regarded as certain that the ways and means so far discovered and reported for effecting things and developing products are mostly rather petty affairs; all major power is dependent on the forms, and derived in order from them; they are the sources, and none of them has yet been discovered.

And therefore (we have said elsewhere) no one reflecting on the machines and battering rams of the ancients would come up with the invention of the cannon operating by means of gunpowder, however persistently he worked at it, even if he spent his life on it. Nor again would anyone who fixed his thoughts and observations on wool-working and plant-derived threads[41] ever have discovered by this means the nature of the silkworm or of the silk derived from it.

---

[41] Presumably linen or cotton

And so (when you think about it) all that can be regarded as the more remarkable inventions have come to light quite by chance, and not through little refinements and extensions of arts. Nothing exhibits them more quickly,[42] moves faster, than chance (whose way of achievement takes centuries), except the discovery of forms.

There is no need to give particular examples of such instances, because there are so many of them. What is absolutely needed is to do a thorough survey and examination of all the mechanical arts, and of the liberal arts too (so far as they may result in works), and then to make a compilation or particular history of the great accomplishments, the magisterial achievements and finished works, in each art, together with their modes of working or operation.

We do not however limit the hard work which needs to be done in making a compilation of this kind to rehearsing the acknowledged masterpieces and secrets of an art which arouse our wonder. For wonder is the child of rarity; since rare things cause wonder even though they are in their kind of common natures.

On the other hand, things which are truly wonderful because of their difference in kind compared with other species are barely noticed if they are in common use around us. We should take note of *unique* instances of art, as well as *unique* phenomena of nature, as we have said before. Just as we included the sun, the moon, the magnet and so on among the *unique instances* of nature, though they are very ordinary albeit of almost unique nature, so we should do the same for the *unique instances* of art.

For example: paper is a *unique instance* of art; a common enough thing. But look at the matter carefully. Artificial materials are either simply woven from upright and horizontal threads, like cloth made of silk, wool, linen etc.; or made of dried liquids, such as brick, or earthenware, or glass, or enamel, or porcelain, and so on; they shine if closely compacted; if not, they harden but do not shine. Now everything made of dried liquids is fragile, and not in the least tacky or tenacious. Paper however is a tenacious substance which can be cut and torn; so that it imitates and almost rivals the skin or membrane of an animal, or the foliage of a vegetable, and such natural products. It is not fragile like glass nor woven like cloth; it certainly has fibres, but not distinct threads, altogether like natural materials. And

---

[42] *repraesentat*: ef. I.109 ad fin.

so it is found to have hardly any similarity with other artificial materials but to be absolutely unique. The better kinds of artificial materials are surely those which either most closely imitate nature, or on the other hand masterfully rule her and change her completely.

Again, among the *contrivances and tools of man,* we should not condemn tricks and toys out of hand. Their applications are trivial and frivolous, but some of them may be useful for information.

Finally, we must not altogether ignore superstition and (as long as the word is taken in its common sense) magic. Such things are deeply buried under great heaps of fables and lies, but one should still look into them a little to see whether perhaps some natural operation lives a latent life in any of them; as in spells, the fortification of the imagination, agreement between distant objects, the transmission of impressions from spirit to spirit no less than from body to body, and so on.

XXXII

It is obvious from what has been said that instances of the last five kinds discussed (that is, *instances of resemblance, unique instances, deviant instances, borderline instances and instances of power*) should not be kept back until we are investigating a particular nature, in the way in which we should keep back the instances we listed first and some of the following ones. Rather we should make a collection of them, a kind of particular history, right at the start; so that they may organise what enters the intellect, and improve the intellect's own feeble condition, which simply cannot avoid being impressed and coloured, and in the end twisted and distorted, by the constant assault of everyday impressions.

We should employ these instances as a preliminary training, to correct and cleanse the intellect. Anything that draws the intellect away from common things smooths and polishes its surface for the reception of the clear, dry light of true ideas.

Such instances also lay and pave the road that leads to practical application, as we shall argue at the appropriate point when we discuss *deduction to practice.*[43]

---

[43] Bacon never wrote this proposed section of *The New Organon.*

XXXIII

As the eleventh of the *privileged instances* we shall put *instances of association* and *of aversion*,[44] which we have also chosen to call *instances of unchanging propositions*. These are the instances which exhibit a substance or concrete thing, in which the nature under investigation either inevitably follows, like an indivisible companion, or on the other hand always turns and flees and is excluded from association, like an enemy and an alien. On the basis of such instances we form sure, universal propositions, either affirmative or negative, in which the subject will be a particular body in concrete form, and the predicate the sought nature. For particular propositions are not *unchanging* at all; they are propositions where the sought nature is found to be fluid and unstable in a concrete object; i.e. it is a nature which is approaching or acquired, or on the other hand departing or shed. Thus particular propositions have no great privilege, except only in the case of *transition* which we discussed above.[45] Nevertheless, even particular propositions are valuable when collated and compared with universal propositions, as we shall argue at the appropriate point. But even in universal propositions we do not require total or absolute affirmation or negation. It is adequate for the purpose if they admit some unique or rare exception.

Thus the function of *instances of association* is to narrow the scope of the *affirmative* of a form. In *instances of transition* the *affirmative* of a form is narrowed down, so that we have to understand the form of the thing as something which is accepted or rejected by the act of *transition*. Similarly in *instances of association* the *affirmative* of a form is narrowed down, so that we are compelled to recognise the form of the thing as something which enters into the composition of such a body, or on the other hand refuses to do so; so that anyone who becomes familiar with the structure or figure of such a body is close to bringing the form of the sought nature into the light.

For example: suppose the sought nature is heat. The *associated instance* is flame. For in water, air, stone, metal and most other things heat is mobile, and can come and go; but all flame is hot, so that heat is always associated with the formation of flame. No *instance of aversion* to heat occurs in our experience. We have no sense experience of the bowels of the earth; but

---

[44] *instantiae comitatus, atque hostiles*
[45] II.23.

there is no concretion of bodies known to us which is not susceptible of heat.

Or suppose that the sought nature is Solidity. An *averse instance* is air. Metal may be molten or solid; so may glass; even water can be solid, when it freezes: but it is impossible for air ever to solidify, or lose its fluidity.

There remain two cautions about *instances of unchanging propositions* relevant to our discussion. First if there is no absolute universal, *affirmative* or *negative*, we should carefully note that fact itself as non-existent, just as we did in the case of heat, where the universal negative (so far as entities within our experience are concerned) is lacking in nature. Similarly, if the nature under investigation is eternal or incorruptible, the affirmative universal is not available in our experience. For *eternal* or *incorruptible* cannot be predicated of any substance beneath the heavens, or above the interior of the earth. The second caution is that universal propositions, both negative and positive, about a concrete thing have concrete things attached to them which seem to approach the non-existent. In the case of heat, very gentle, barely burning flames; in the case of incorruptibility, gold, which is the nearest thing to it. All these things point to the natural boundaries between being and non-being; and help to define the limits of forms, so that they do not swell and stray beyond the conditions of matter.

XXXIV

Twelfth among privileged instances we shall put those *accessory*[46] *instances*[47] of which we spoke in the previous aphorism; which we have also chosen to call *instances of the end* or *terminal instances*. Such instances are not only useful when attached to unchanging propositions, but also in themselves and of their own nature. For they clearly mark the true divisions of nature, the measures of things, the vital *how far* a nature can do or suffer anything, and the transition from one nature to another. Such are, in weight, gold; in hardness, iron; the whale in animal size; the dog in smell; the lighting of gunpowder in swiftness of expansion; and so on. They show the final degrees at the bottom of the scale no less than at the top; as spirit of wine in weight; silk in softness; skin grubs in animal size; and so on.

[46] *Subjunctiva* resumes the grammatical analogy: 'subjunctive' mood.
[47] *instantiae subjunctivae*

XXXV

Thirteenth among *privileged instances* we shall put *instances of alliance* or *of union*.[48] They are instances which fuse and unite natures thought to be heterogeneous, and are noted and marked as such in the accepted distinctions.

Instances of alliance reveal that the operations and effects which are classed as belonging to one heterogeneous nature also belong to other heterogeneous natures; and this proves that the supposed heterogeneity is not genuine or essential, but simply a modification of a common nature. They are therefore of the highest value in lifting and raising the mind from differences to *genera,* and for getting rid of the phantoms and false images[49] of things as they come forward to meet us in disguise in concrete substances.

For example: suppose the nature under investigation is Heat. There seems to be a quite usual and authentic distinction between three kinds of heat, viz. the heat of the heavenly bodies, the heat of animals and the heat of fire; and these heats (especially one of them compared to the other two) are different and quite heterogeneous in their actual essence and species, or in their specific nature; for the heat of heavenly bodies and of animals creates and nourishes, whereas the heat of fire corrupts and destroys. An *instance of alliance* therefore is the quite common experience of bringing a vine branch into a house where a fire is kept burning on the hearth, so that the grapes ripen as much as a month earlier than outside; so that one can speed the ripening of the fruit, even when it is hanging on the tree, by using a fire, though this seems to be the proper function of the sun. From this beginning the mind rejects essential heterogeneity, and readily rises to inquire what real differences are found between the heat of the sun and fire which make their operations so dissimilar, despite their sharing a common nature.

The differences will be found to be four: first, the heat of the sun by comparison with the heat of fire is much gentler and softer in degree. Secondly (particularly as it reaches us through the air), it is much more humid in quality. Thirdly (and this is the main difference), it is supremely inconstant, now approaching and increasing and then receding and diminishing, which is what most contributes to the generation of bodies. For Aristotle rightly asserted that the principal cause why things come to be and pass

[48] *instantiae foederis sive unionis*
[49] 'phantoms and false images' (Ellis)

away, here on the surface of the earth, is the oblique path of the sun through the zodiac; as a result of which the sun's heat is actually amazingly inconstant, partly by the alternation of day and night and partly by the succession of summers and winters. Yet Aristotle never fails immediately to twist and distort what he had truly discovered. For in his habitual role as arbiter of nature he magisterially determines that the approach of the sun is the cause of coming-to-be and its recession the cause of passing away; whereas both things (i.e. the sun's approach and the sun's recession) are a reason for both coming-to-be and passing away indifferently, not respectively, since inequality of heat causes things to be born and to pass away, while equality of heat favours only preservation. There is also a fourth difference between the heat of the sun and the heat of fire, a difference of immense significance: viz. that the sun deploys its operations through long spaces of time, whereas the operations of fire (under pressure from man's impatience) are brought to a result in a relatively brief period. However, one might take constant care to control the heat of a fire and reduce it to a mild and moderate temperature (there are many ways to do this), and one might also sprinkle and mingle moisture with it, and in particular one might imitate the inconstancy of the sun, and finally be patient and accept the time it takes (this would certainly not be a period like that of the works of the sun, but still a longer time than men normally take in operations with fire). If one does all this, he will easily dismiss the [notion of] the heterogeneity of heat, and will with the heat of a fire either approach or equal or in some cases surpass the operations of the sun. A similar *instance of alliance* is the resuscitation, with a little warmth from a fire, of butterflies stunned and half-killed by cold; by which you may easily see that fire is no more barred from reviving animals than from ripening vegetables. Also Fracastoro's[50] famous invention of the highly heated pan with which doctors cup the heads of stroke victims whose lives are despaired of, obviously expands the animal spirits which are suppressed and almost extinguished by humours and obstructions of the brain; and stirs them into activity; it works as fire works on water or air, and yet it has the effect of restoring life. Eggs too are sometimes hatched by the heat of a fire, in direct imitation of animal heat; there are several other such instances; so that no one can doubt that the heat of fire can, in many subjects, be tempered to be an image of the heat of heavenly bodies and animals.

---

[50] Girolamo Fracastoro (1483–1553), physician and poet of Verona and author of the poem *Syphilis* (1530).

Similarly, suppose the sought natures are motion and rest. There seems to be a common division, which is also drawn from the heart of philosophy, that natural bodies either rotate, or move in a straight line, or stand and stay at rest. For there is either motion without end, or rest at an end, or movement towards an end. Perpetual motion of rotation seems to be peculiar to heavenly bodies; staying or rest seems to belong to the globe of the earth itself; but other bodies, which they call heavy and light, that is bodies which are outside their natural places, move in a straight line towards masses or accumulations of things like themselves: light things upwards towards the circuit of heaven, heavy things downwards towards the earth. These are fine things to say.

A low comet is an *instance of alliance;* though it is far below the heaven, it rotates. And the fiction of Aristotle about a comet's attachment to or pursuit of a particular star has long been discredited; not only because the explanation of it is not probable, but because of the observed fact of the wandering and irregular movement of comets through the different areas of the sky.

Again another *instance of alliance* in this subject is the motion of air; which seems itself to revolve from east to west between the tropics (where the circles of rotation are greater).

Yet another *instance* would be the ebb and flow of the sea, provided that the waters themselves were seen to move with a motion of rotation (albeit slow and difficult to observe) from east to west; but in such a way that they are driven back twice a day. If that is the case, it is obvious that the motion of rotation is not confined to the heavens but is shared with air and water.

The ability of light objects to move upwards varies somewhat. Take a bubble of water as an instance of alliance in this case. Air under water ascends rapidly towards the surface because of the motion of a 'blow' (as Democritus[51] calls it) by which descending water strikes and raises the air up, and not by struggle or striving on the part of the air itself. On reaching the water's surface, the air is prevented from rising further by the light resistance which it finds in the water not immediately allowing itself to be parted; and thus the air's own tendency upwards is very slight.

Likewise suppose the sought nature is weight. It is a very well-accepted distinction that dense and solid objects tend towards the centre of the earth,

---

[51] For Democritus see note on I.51.

while light and rare objects tend towards the circuit of the sky, as towards their own places. And as regards places, it is quite silly and childish to think (though thoughts of this kind are prevalent in the schools) that place has any influence at all. Hence philosophers talk nonsense when they say that if a hole were made through the earth, heavy bodies would stop when they came to the earth's centre. For it would surely be a wonderfully powerful and effective kind of nothing or mathematical point which had an effect on other things and which other things would seek; for body is acted on only by body. But the tendency to move up and down is either in the structure of the moving body or in a sympathy or agreement with another body. If any body is found which is dense and solid but which nevertheless does not tend towards the ground, such a distinction is shattered. If Gilbert's opinion[52] is accepted that the magnetic force of the earth to attract heavy objects does not extend beyond the circle of its own power (which operates always up to a certain distance and not beyond it), and this is verified by some instance, it will in fact be an *instance of alliance* in this subject. However, no certain and obvious instance on this point has come to notice up to now. The nearest thing to it seems to be the waterspouts which are often seen on voyages through the Atlantic Ocean to either Indies. So great apparently is the mass and force of water suddenly released by such water-spouts that there seems to have been a prior accumulation of water which has stalled and stayed in these places, and then been thrown and forced down by some violent cause, rather than to have fallen by the natural motion of its weight. Hence it may be conjectured that a dense and compact physical mass at a great distance from the earth will hang in suspense like the earth itself; and will not fall unless thrown down. But we claim no certainty here. And in this and many other cases, it will easily be seen how poor we are in natural history, since we are sometimes compelled to adduce suppositions instead of certain instances.

Likewise suppose the sought nature is the discursive movement of the mind. It seems to be quite correct to make a distinction between human reason and the cleverness of animals. And yet there are some instances of actions done by animals which suggest that animals do seem to go through a chain of reasoning: the story is told of a crow almost dead of thirst in a great drought, who caught a glimpse of some water in the hollow trunk of a tree, and since the trunk was too narrow to get into, it dropped in pebbles

---

[52] For Gilbert see note on I.54.

one after the other so that the water level would rise for it to drink: this became proverbial.[53]

Likewise suppose the sought nature is the visible. There seems to be a perfectly true and certain distinction between light, which is the original visible and the primary source of sight, and colour, which is a secondary visible and is not seen without light, so that it seems to be nothing other than an image or modification of light. And yet there appear to be *instances of alliance* on this on both sides: i.e. snow in large quantities and a sulphur flame; in one of which there appears to be a colour just becoming light, and in the other light verging on colour.

XXXVI

In the fourteenth place among *privileged instances* we place *crucial instances*;[54] we take the term from the *signposts* which are erected at forks in the road to indicate and mark where the different roads go.[55] We have also chosen to call them *decisive instances* and *instances of verdicts*, and in some cases *oracular* and *commanding instances*. This is how they work. Sometimes in the search for a nature the intellect is poised in equilibrium and cannot decide to which of two or (occasionally) more natures it should attribute or assign the cause of the nature under investigation, because many natures habitually occur close together; in these circumstances crucial instances reveal that the fellowship of one of the natures with the nature under investigation is constant and indissoluble, while that of the other is fitful and occasional. This ends the search as the former nature is taken as the cause and the other dismissed and rejected. Thus instances of this kind give the greatest light and the greatest authority; so that a course of interpretation sometimes ends in them and is completed through them. Sometimes crucial instances simply occur, being found among instances long familiar, but for the most part they are new and deliberately and specifically devised and applied; it takes keen and constant diligence to unearth them.

For example, suppose the nature under investigation is the ebb and flow of the sea, repeated twice a day, six hours for each incoming and outgoing tide, with some difference which corresponds to the motions of the moon. The fork of the road in this nature is as follows.

---

[53] Avianus, *Fables*, 27, 'The Crow and the Pitcher'.
[54] *inistantiae crucis*
[55] 'Crucial' is derived from *crux*, 'cross'.

This motion has to be caused either by a forward-and-backward movement of the waters, like water sloshing back and forth in a basin, which leaves one side of the basin when it covers the other, or by the waters rising and subsiding from the depths, like water boiling up and then subsiding. But one is in doubt to which of the two causes to attribute the ebb and flow. If the first account is accepted, then when there is a high tide in the sea on one side, there has to be a low tide in the sea at the same time elsewhere. And so this is the form the investigation will take. And yet Acosta[56] and many others have observed (after careful investigation) that there are high tides at the same time on the shores of Florida and on the opposite shores of Spain and Africa, and likewise low tides at the same times, and not the contrary, that when it is high tide on the shores of Florida, it is low tide on the shores of Spain and Africa. However, if one looks at it still more carefully, this does not prove a rising motion and refute a forward motion. For it can happen that waters move forward, while flooding both shores of a stretch of water at the same time, i.e. if those waters are subjected to force and pressure from another direction, as happens in rivers, where the ebb and flow on both shores occurs at the same time, even though the motion is clearly motion forward, the motion of waters entering the river mouth from the sea. And so it may be similarly that waters coming in a great mass from the East Indian Ocean are driven and thrust into the basin of the Atlantic Sea, and thus flood both sides at the same time. We must therefore ask whether there is another basin through which the waters can at the same time ebb and flow back. And there is, the Southern Sea, which is not smaller than the Atlantic, but rather wider and vaster, and would be adequate for this purpose.

And so we have reached the *crucial instance* on this subject. Here it is: if it is found for certain that when it is high tide on the opposing shores of both Florida and Spain in the Atlantic Sea, there is at the same time high tide on the shores of Peru and near the mainland of China in the Southern Sea; then by this *decisive instance* we must certainly reject the assertion that the ebb and flow of the sea (the subject of the inquiry) occurs by a forward motion; there is no other sea or place remaining where there could be a retreat or ebb at the same time. This could most conveniently be known if inquiry were made of the inhabitants of Panama and Lima (where the two Oceans, the Atlantic and the Southern, are separated by a small Isthmus)

[56] Jose de Acosta (*c.* 1539–1600), Spanish Jesuit missionary to Peru. Published *The Natural and Moral History of the Indies* (1590; translated into English 1604).

whether the ebb and flow of the sea on the two sides of the Isthmus occur at the same time or not. Now this verdict or definitive rejection seems certain if we posit that the earth is unmoving. However, if the earth rotates, it may perhaps be that the earth and the waters of the sea rotate unequally (unequal in speed or force), the consequence of which would be violent pressure forcing the waters up into a heap, which is the high tide, and a subsequent dropping of the waters (when they can no longer stay heaped up), which is the ebb. This needs a separate inquiry. But on this supposition also it remains equally true that there has to be an ebb of the sea somewhere at the same time as there are high tides elsewhere.

Likewise suppose the nature under investigation is the latter of the two motions which we first assumed, viz. arising and subsiding motion of the sea if, after careful examination, we do in fact reject the other motion we mentioned, the forward motion. There will then be a three-way fork in the road: the motion by which waters rise and fall in their ebbs and flows, without the addition of other waters flowing into them, has to occur in one of three ways. It may be because a great mass of water wells up from the interior of the earth and then sinks back into it; or because there is not a larger amount of water, but the same waters (with no increase of quantity) are stretched or thinned out so that they occupy a greater space and dimension, and then contract; or because the quantity and extension are no greater, but the same waters (same both in quantity and in density and rarity) rise and fall through some magnetic force above which attracts and draws them up, through agreement.[57] Let us leave aside the first two motions and narrow the inquiry (if you please) to this last [possibility]; and let inquiry be made whether there is any such raising by agreement or magnetic force. And first it is evident that all the waters, as they lie in the trench or bed of the sea, cannot be raised together at the same time, because there would be nothing to take their place at the bottom; hence if the waters had any such tendency to rise, it would be broken and restrained by the bonds of nature, or (as they commonly say) to prevent the occurrence of a vacuum. The only explanation left is that the waters rise in one place and, for that reason, fall and recede in another place. In fact, it will necessarily follow that since the magnetic force cannot operate over the whole, it operates most intensely over the centre, so that it lifts the waters in the middle, and when they are raised, they move away for the sides and leave them bare and uncovered.

---

[57] 'Agreement' = *consensus*. Bacon contrasts *consensus* with *sympathia* (a term he rejects) at 11.50(6).

And thus at last we have reached the *crucial instance* on this subject. It is this: if it is found that in the ebbs of the sea the surface of the waters in the sea is more arched and rounded as the waters rise in the middle of the sea and fall away at the edges, which are the shores; and in the flows the same surface is planer and flatter when the waters return to their former position; then by this *decisive instance* we can certainly accept raising by magnetic force; otherwise it has to be totally rejected. And this is not difficult to find out by making use of sounding lines in straits; i.e. to find out whether towards the centre of the sea the water is higher or deeper in ebbs than in flows. And we must note that if this is so, the fact (contrary to belief) is that the waters rise in ebbs, and fall only in flows, so as to cover and flood the beaches.

Similarly, suppose the nature under investigation is the spontaneous motion of rotation, and particularly whether the diurnal motion by which the sun and the stars rise and fall to our view is a true motion of rotation in the heavens, or an apparent motion in the heavens but a true one in the earth. A *crucial instance* on this subject would be as follows. If we find in the ocean a motion from east to west, however feeble and sluggish; if we find the same motion a little more rapid in the air, especially within the tropics, where it is easier to detect because of the larger circumference; if we find the same motion in the lower comets, now in a strong and lively form; if we find the same motion in the planets, yet so apportioned and graduated that the shorter the distance from the earth, the slower it is, the further the faster, and at its swiftest in the heaven of the stars: then we should certainly acknowledge the truth of diurnal motion in the heavens, and must deny the motion of the earth; because it will be clear that the motion from east to west goes throughout the cosmos, and is based on the agreement of the whole universe, and is at its most rapid in the heights of heaven, and grows faint by degrees, and finally languishes and fails at the unmoving point, namely the earth.

Similarly, let the nature under investigation be the other motion of rotation which the astronomers often cite, which resists and runs contrary to the diurnal motion, namely motion from west to east; which the old astronomers attribute to the planets and to the heaven of the stars as well, but Copernicus[58] and his disciples attribute also to the earth; and let it be asked whether any such motion is found in nature, or whether it is rather

---

[58] Nicolaus Copernicus (1473–1543) published *De Revolutionibus orbium coelestium* in 1543. It includes a heliocentric account of the solar system.

a fiction and a supposition to speed and shorten calculations, and to support that pretty theory which explains the heavenly motions by perfect circles. For this motion is certainly not shown to be a true and real motion in the heavens either by the fact that a planet fails to return to the same point of the starry heaven in its daily motion, or by the difference between the poles of the zodiac and the poles of the earth; which are the two things that have encouraged the idea of this motion. For the first phenomenon is best saved by precedence and dereliction;[59] and the second by spiral lines; so that the inequality of return and the decline towards the tropics may be rather modifications of the one true diurnal motion than resistant motions or motions around different poles. And it is quite certain, if we take the view for the moment of the common man (dismissing the fictions of astronomers and schoolmen, who have the habit of making unwarranted assaults on common sense, and preferring obscurities), that to the senses the motion is such as we have described it; whose image we once represented (as in a machine) by iron wires.

But a *crucial instance* in this subject could be as follows. If we found in an account worthy of belief; that there had been a comet, either higher or lower, which did not rotate in evident agreement (even if rather irregular) with the diurnal motion, but rather rotated against the direction of the heavens; then certainly this much has to be allowed, that there may be some such motion in nature. But if we find nothing of the kind, we must regard it as suspect, and have recourse to other *crucial instances* on the point.

Similarly, suppose the nature under investigation is weight or heaviness. The fork in the road on this nature is as follows. Heavy and weighty things must necessarily either tend of their own nature towards the centre of the earth because of their own structure; or be attracted and drawn by the physical mass of the earth itself as by an agglomeration of connatural bodies, borne towards it by agreement. But if the latter is the reason, it follows that the closer heavy things get to the earth, the more powerfully they are borne towards it, and the greater their impetus; the further they are from it, the more feebly and slowly (as is the case with magnetic attraction); and that this occurs within a certain space; so that if they are so distant from the earth that its power cannot act on them, they will remain in suspense, like the earth itself; and will never fall.

And therefore a *crucial instance* on this matter could be as follows. Take

---

[59] 'by supposing that the fixed stars outrun the planets, and leave them behind' (Ellis)

one of those clocks which move by lead weights, and one of those which move by a compressed iron spring; and let them be accurately tested, so that neither of them is faster or slower than the other; then let the clock that moves by weights be placed on the top of a very high church, the other kept below; and let it be noted whether the higher clock moves more slowly than it did because the weights have less power. Let the same experiment be done at the bottom of mines deep below the earth, to see whether a clock of this kind does not move faster than it did, because the weights have increased power. If it is found that the power of the weights decreases at a height and increases under the earth, attraction from the physical mass of the earth may be taken as the cause of weight.

Similarly, let the nature under investigation be the polarity of an iron needle touched by a magnet. The fork in the road for such a nature will be as follows. The touch of a magnet must necessarily either confer polarity towards north and south on the needle of itself; or merely excite the iron and make it ready, the motion being conferred by the presence of the earth, as Gilbert thinks and tries so hard to prove. And thus all the cases which he has collected with perceptive diligence come down to this: namely that an iron nail, which has long lain in a north–south position, gathers polarity by the long lapse of time, without being touched by a magnet; as if the earth itself; which works weakly because of the distance (for the surface or outer crust of the earth is without magnetic virtue, according to him), does nevertheless, given enough time, act like the touch of a magnet, and excite the iron and conform it to itself in its excited state and turn it. Again, his work shows that if a piece of iron, heated red-hot,[60] lay, as it went cold, in a north–south position, it too would gather polarity without the touch of a magnet: as if[61] the parts of the iron which were set in motion by being heated and afterwards contracted in the very moment of their getting cold, were more susceptible and sensitive to the force emanating from the earth than at other times, and hence were as it were excited. But though these things are well observed, they do not prove quite what he claims.

A *crucial instance* on this subject would be as follows: take a magnetic compass and mark its poles; and let the poles of the compass be placed east–west, not north–south; and let them so lie; then place an untouched iron needle on it and allow it to remain for six or seven days. While the needle remains over the magnet (of this there is no doubt) it will ignore

[60] Reading *candens* for *cadens*. Kitchin clearly did the same, translating 'ignited and red-hot'.
[61] Reading *ac* for *sc*.

the poles of the earth and align itself with the poles of the magnet; and thus as long as it remains like this, it is turned towards the east and west of the world. But when we remove the needle from the magnet and set it on a pivot, if we find it immediately turning north–south, or even moving gradually in that direction, then we must accept the presence of the earth as the cause. But if it either turns (as it did before) to the east and west or loses its polarity, we must treat that reason as suspect, and make further inquiry.

Similarly, let the nature sought be the physical substance of the moon: whether it is light, fiery or airy, as most of the old philosophers believed, or solid and dense, as Gilbert and many of the moderns hold, together with some of the ancients. The reasons for the latter view are based especially on the fact that the moon reflects the rays of the sun; and it seems that reflection of light occurs only from solids.

And thus the *crucial instances* on this subject will be those (if there are any) which show reflection from a light body, such as flame so long as it is sufficiently dense. Certainly one cause of twilight, among others, is the reflection of the rays of the sun from the upper part of the air. We also sometimes see the rays of the sun reflected, on fine evenings, from wisps of rosy cloud, with no less brightness, in fact with a clearer and more glorious brightness than that given off by the body of the moon; but it is not established that those clouds have formed into a dense body of water. We also see at night that dark air reflects the light of a candle in a window as well as a dense body would. The experiment should be tried of letting the rays of the sun pierce through a hole on to a dusky blue flame. Certainly when the rays of the full sun fall on not very bright flames, they seem to kill them, so that they look more like white smoke than flames. These are the things that occur to me as examples of *crucial instances* in this matter; and probably better ones can be found. But one must always bear in mind that no reflection is to be expected from a flame unless the flame has a certain depth; otherwise it verges on the transparent. But what should be regarded as certain is that light on an even body is always either taken and passed through, or reflected.

Similarly, let the nature under investigation be the motion through the air of missiles, e.g. javelins, arrows, balls. The School (in its usual manner) deals with this very carelessly, being satisfied to distinguish it under the name of violent motion from what they call natural motion; and as for the first blow or thrust, is content with saying 'that two bodies cannot be in one place, lest there be a penetration of dimensions', and takes no account of the further progress of this motion. But the fork of the road in this matter is as follows: the motion is caused either because the air carries the body which is being propelled, and gathers behind it, as a river does to a boat or a wind to straws; or because the parts of the body itself do not resist the blow, but run forward one after the other to get away from it. Fracastoro[62] accepts the first view, and so do nearly all those who have made more than a cursory investigation of this motion; and there is no doubt that the air plays some role in this; but the other motion is undoubtedly the true one, as an infinite number of experiments prove. One *crucial instance* among others on this subject is as follows: flex a sheet of iron or a strong piece of iron wire, or a reed or quill pen sliced down the middle, and bend it between finger and thumb, and it will spring away. It is obvious that this cannot be attributed to the air gathering behind the body, because the source of motion is in the middle of the sheet or reed, not at the ends.

Similarly, let the nature under investigation be the rapid, powerful expansive motion of gunpowder into flame, which destroys such massive objects and shoots such immense weights as we see in mines and cannon. The fork in this nature is as follows. Either the motion is initiated by the simple tendency of the expanding body when it is set on fire, or in addition by the tendency of the crude spirit, which rapidly flees the fire, and as it pours around it, breaks violently away, like a horse from the starting gate. But the School and common opinion deal only with the first tendency. For men think that it is a fine piece of reasoning to assert that, because of the form of its element, a flame is endowed with a kind of necessity to occupy a larger space than the same body had filled when in the form of powder, and that is why the motion follows. However, they fail to notice that though this is true (since the flame is actually generated), still its generation may be prevented by a mass of stuff stifling and suffocating it, so that the process does not reach the necessity they speak of. For they are right in thinking that if flame is generated, expansion must occur and emission or ejection of the body obstructing it must follow. But that necessity is obviously averted if the solid mass suppressed the flame before it was generated. And we see that a flame, especially at its first generation, is soft and gentle, and requires a space in which it can try itself out and play. And therefore we cannot ascribe so much force to the thing in itself. But this is true: the

---

[62] For Fracastoro see n. 50 above.

generation of these blasting flames and fiery winds happens as the result of a struggle between two substances whose natures are totally contradictory to each other. One is highly inflammable, the nature which flourishes in sulphur; the other abhorring fire, as does the crude spirit in nitre. The result is a marvellous conflict as the sulphur fires itself so much as it can (for the third substance, namely willow charcoal, does little more than join the other two substances and properly unite them), and the spirit of nitre breaks out, so far as it can, and at the same time expands (for air too and all crude substances, as well as water, react to heat by expanding), and as it flees and bursts out, it blows the flame of sulphur in all directions, like a hidden bellows.

There could be two kinds of *crucial instances* in this subject. The one consists of substances which are most inflammable, such as sulphur, camphor, naphtha and so on, with their mixtures; which take fire more swiftly and easily than gunpowder if they are not obstructed; and this makes it clear that it is not the simple tendency to catch fire in itself which has that stupendous effect. The other consists of substances which avoid fire and shrink from it, such as all the salts. For we see that if they are thrown into a fire, a watery spirit breaks out, with a crackling noise, before fire starts; and this also happens in a gentler way even with leaves if they are a little bit resistant, as the watery part breaks out before the oily part catches fire. But it is seen best in quicksilver, which is well called mineral water. For apart from catching fire, it almost equals the power of gunpowder in sheer eruption and expansion; and when it is mixed with gunpowder it is said to compound its strength.

Similarly, let the nature under investigation be the transitory nature of flame and its instant extinction. For the nature of flame does not appear to be fixed and stable here on earth, but to be generated at every moment and directly extinguished. For it is evident that in the case of flames, in our experience, which last and endure, their duration is not that of the same flame as an individual, but is maintained by a succession of new flames constantly generated, and the numerically identical flame lasts only a very short while; and this is easily seen from the fact that the flame dies as soon as you take away its fuel or nourishment. The fork in this matter is as follows. The momentary nature occurs either because the cause which first generated it lets up, as in the case of light, sounds and the so-called violent motions; or because the flame could in its nature endure here on earth, but suffers violence from contrary natures around it, and is destroyed.

Therefore a *crucial instance* on this matter may be as follows. In great fires we see how high flames leap. For the wider the base of the flame, the higher the top. And so it seems that extinction begins to occur at the edges, where the flame is oppressed by air and is feeble. But the central parts of the flame, which the air does not touch and which is surrounded by other flame on all sides, remain numerically the same and are not extinguished until they are squeezed in bit by bit by the air around the edges. And that is why all flame is pyramidal, wider at the base near the fuel and more pointed at the top (where the air threatens it and the fuel cannot keep it supplied). Smoke, the other hand, which is narrower at the base, gets wider as it rises, and becomes like an inverted pyramid; that is because air accepts smoke, and compresses flame (nor should anyone imagine that air is burning flame, since they are completely heterogeneous substances).

There could be a more precisely appropriate *crucial instance* on this matter if the thing could by chance be exhibited through two-coloured flames. So let a small metal candle-holder be taken, and a small lighted wax candle fixed in it; let the candle-holder be placed in a broad shallow bowl, and a moderate amount of spirit of wine be poured around it, but not enough to reach the lip of the candle-holder; then light the spirit of wine. The spirit of wine will give a bluish flame, while the candle lamp gives a yellower one. Notice then whether the lamp flame (which it is easy to distinguish from the flame of the spirit of wine by the colour; for flames, like liquids, do not mingle instantly) remains pyramidal, or whether rather it tends more to the shape of a globe, since there is nothing there to destroy or compress it. And if the latter occurs, it is to be laid down as certain that the flame remains numerically identical so long as it is wrapped up in the other flame.

And so much for *crucial instances*. We have deliberately taken quite a long time to deal with them, so that men may gradually learn the habit of forming judgements of a nature by *crucial instances* and illuminating experiments, and not by probable reasoning.

XXXVII

As fifteenth of the privileged instances I will put *instances of divergence*[63] which indicate separations of natures which most commonly occur.[64] They

---

[63] *instantiae divortii*
[64] or possibly 'occur together'

differ from the instances appended to the *instances of association*[65] because they declare the separation of a nature from a concrete object with which it is normally found, while the latter point to the separation of one nature from another. They are also different from *crucial instances;* because they settle nothing, but merely point to a separation of one nature from another. Their value lies in showing up false forms, and in making short work of hasty reflections inspired by passing objects; so that, we may say, they add lead and weights to the intellect.[66]

For example: let the natures under investigation be the four natures which Telesio[67] regards as *house companions* and, one might say, from the same room, viz. heat, brightness, subtlety, mobility or readiness for motion. Many, many instances of separation are found among them. For air is subtle and apt to move, but not hot or bright; the moon is bright, without heat; boiling water hot, without light; the motion of the iron needle on a pivot is quivering and agile, though in a substance which is cold, dense and opaque; there are quite a few other examples.

Similarly, let the natures under investigation be physical body and natural action. Natural action seems only to occur when existing in a body. Even so there might perhaps be an *instance of divergence* in this matter. This is the magnetic action by which iron is drawn towards a magnet, heavy objects towards the globe of the earth. We might add some other operations at a distance. For such action happens both in time, at intervals and not at one point of time, and in space, by degrees and through distances. There is therefore some moment of time and some interval of space in which the force or action hangs in the midst between the two bodies which are causing the motion. Hence our reflection is focused on this question: whether the bodies which are the terminals of the motion affect or alter the bodies between, so that the force moves from one terminal to the other by succession and true contact, and for a time subsists in the body between; or is there nothing there but bodies and forces and spaces? And in optical rays and sounds and heat and some other things that work at a distance, it is probable that the bodies in between are affected and altered: the more so as this requires a medium suited to carrying such an operation. But magnetic or connective force is indifferent as to medium, and the force is not impeded in any kind of medium. But if the force or action has nothing in

[65] See 33 or 34 above.
[66] For this image, cf. 1.104.
[67] For Telesio see 1.116.

common with the intervening body, it follows that it is a natural force or action subsisting for some time in some space without a body, since it is not subsisting either in the terminals or in the medium. And thus magnetic action may be an *instance of divergence* in a physical substance and a natural action. Something should be added to this as a corollary or benefit not to be missed: that even in philosophising on the basis of the senses, one may have proof that there are separate and incorporeal entities and substances. For if the natural force and action which emanate from a body could subsist in some time and place wholly without a body, it is also close to being able to emanate in its origin from an incorporeal substance. For physical substance seems no less required for sustaining and carrying natural action than for initiating or generating it.

XXXVIII

There follow five sets of *instances* which we have chosen to call in one general word *instances of the lamp* or *of first information*. They are those which assist the senses. Since all interpretation of nature begins from the senses, and runs by a straight, even, well-made road from perceptions of the senses to perceptions of the intellect, which are true notions and axioms, it is necessary that the fuller and more exact the presentations or exhibits of the senses themselves are, the more easily and successfully everything will go.

Of these five *instances of the lamp*, the first set strengthen, enlarge and correct the immediate actions of sense; the second make the non-sensible sensible; the third point to continuous processes or series of things and motions which (in general) are only noticed at their departure or in their high points; the fourth provide a substitute for the senses where the senses can do nothing at all; the fifth attract the attention and observation of the senses, and at the same time limit the subtlety of things. And now we must speak of each one individually.

XXXIX

As sixteenth of the privileged instances we shall place *instances that open doors* or *gates*;[68] this is the name we give to those instances that assist direct

---

[68] *instantiae januae sive portae*

actions of sense. It is evident that sight holds first place among the senses, as far as information is concerned; and so this is the sense for which we must first find aids. There appear to be three kinds of aids: either to see what has not been seen; or to see further; or to see more accurately and distinctly.

Apart from spectacles and such things, whose function is simply to correct and alleviate the weakness of impaired vision, and so provide no new information, an instance of the first kind are microscopes, lately invented, which (by remarkably increasing the size of the specimens) reveal the hidden, invisible small parts of bodies, and their latent structures and motions. By their means the exact shape and features of the body in the flea, the fly and worms are viewed, as well as colours and motions not previously visible, to our great amazement. Furthermore, they say that a straight line drawn by a pen or pencil is seen through such magnifying glasses to be very uneven and wavy; evidently, because neither the movements of the hand, even when assisted by a ruler, nor the impression of the ink or of the colour are really true, though the irregularities are so small that they cannot be seen without the aid of such microscopes. In this matter too men have provided a kind of superstitious commentary (as usual with new and strange matters), viz. that such microscopes illustrate works of nature but discredit works of art. But this is simply because natural textures are much more subtle than artificial textures. For this microscope is only good for tiny things; if Democritus had seen a magnifying glass, he would perhaps have jumped for joy, thinking a means of viewing the atom (which he affirmed was quite unseeable) had been invented. But the inadequacy of such microscopes except for tiny things (and not for tiny things if they are in a larger body) destroys the use of the thing. For if the invention could be extended to larger bodies, or small bits of larger bodies, so that we could see the texture of linen cloth as a net, and in this way make out the tiny, hidden features and irregularities of gems, liquids, urine, blood, wounds and many other things, we could undoubtedly derive great benefits from this invention.

Of the second kind is the other magnifying glass, Galileo's great achievement,[69] telescope, with whose help we may open up and practise a closer approach to the stars, as if by ferries or dinghies. For it establishes that the galaxy is a knot or heap of small stars, which are plainly separate and distinct,

---

[69] Galileo Galilei (1564–1642) of Padua. Bacon gives him the credit for the invention of the telescope; Galileo presented one to the doge of Venice in 1609.

of which the ancients had only a suspicion. It also appears to demonstrate that the spaces between the so-called orbits of the planets are not wholly empty of other stars, but that the heaven begins to be starry before you get to the starry heaven itself; though with smaller stars than can be seen without this telescope. With it one may view the choirs of small stars around the planet Jupiter (and may conjecture from this that there is more than one centre in the motions of the stars). With it the irregularities in the light and dark areas of the moon are more distinctly seen and located; so that a kind of map of the moon may be made. With it one may see the spots in the sun, and things of that kind: all certainly noble discoveries, so far as one may safely credit such demonstrations. But we are very suspicious of such things, because experience stops with these few things, and not many other things which equally deserve investigation have been discovered by the same means.

Of the third kind are rods for measuring the earth, astrolabes and so on, which do not enlarge the sense of sight, but correct and focus it. If there are other instances which assist the other senses in their direct, individual actions, still they do not contribute to our project unless they are such as to add to the actual stock of information which we now have. And so I have not mentioned them.

XL

In the seventeenth place among the *privileged instances* we shall put *summoning instances*,[70] a word we take from the civil courts, because they summon things to present themselves which have not previously done so; we have also chosen to call them *citing instances*. They make sensible the non-sensible.

Things escape the senses either because the object is placed at a distance, or because the senses are obstructed by bodies between themselves and the object, or because the objects are not capable of making an impression on the senses, or because the quantity of the object is not sufficient to strike the senses, or because the time is insufficient to activate the senses, or because the senses cannot stand the effect of the object, or because an object has previously filled and possessed the senses so that there is no room for another motion. These factors pertain primarily to sight, and secondly to

[70] *instantiae citantes*

touch. These two senses are informative in a broad way about ordinary objects; whereas the other three barely give any information except directly and about objects peculiar to each sense.

(1) In the first case, a thing is conveyed to the senses only if the object that cannot be seen has something added or substituted for it which can alert and impress the senses from a distance: such as in carrying news by means of fires, bells and so on.

(2) In the second case, conveyance takes place when things which are kept hidden by the obstruction of a body, and cannot be easily exhibited, are brought before the senses by things which are on the surface, or which come out from inside: as the condition of a human body is revealed by the pulse, the urine and so on.

(3, 4) Conveyances of the third and fourth kinds apply to many objects; in the inquiry into nature we must always be alert for them. Here are some examples. It is evident that air and spirit and things of that kind which are tenuous and subtle throughout cannot be seen or touched. It is absolutely necessary to make use of conveyances in investigating such substances.

Suppose the nature under investigation is the action and motion of spirit enclosed in a tangible body. For every tangible body on earth contains an invisible and intangible spirit; the body envelops and clothes it. This is the powerful source of three effects, the marvellous process of spirit in a tangible body: when the spirit within a tangible thing is released, it contracts and dries up the body; when it is kept in, it softens and melts them; and when it is neither completely released nor wholly kept in, it shapes them, gives them limbs, absorbs, consumes, organises, and so on. All this is conveyed to the senses by the visible effects.

For in every tangible and inanimate body, enclosed spirit first multiplies and feeds on the tangible parts which are most ready and available, and digests and dissolves them, and turns them into spirit, and then they escape together. This multiplication and dissolution by spirit is conveyed to the senses by loss of weight. For when anything dries up, something is lost from its quantity; and the loss is not so much from the spirit which was previously in it, but from the substance which formerly was tangible and has just been converted; for spirit has no weight. The exit or release of spirit is conveyed to sense in the rusting of metals and other kinds of decomposition which stop before they reach the rudiments of life; for the latter belong to the third type of process. For in compact substances the spirit finds no pores or pathways to get out by, and is therefore compelled to force tangible parts out and push them before it, so that they come out with it; and that is how rust and such things occur. Contraction of tangible parts after release of some part of the spirit (which is followed by drying up) is conveyed to the sense by an increase in the hardness of the thing, but much more by the consequent splitting, shrinking, wrinkling and folding in bodies. Bits of wood shrivel and shrink; skins wrinkle; and not only that, but (after a sudden release of spirit by the heat of a fire) they are so keen to contract that they wind around themselves and curl up.

By contrast, when the spirit is kept in while being expanded and stimulated by heat or something analogous (as in the case of substances which are quite solid or tenacious), then the bodies are softened, like red-hot iron; they flow, like metals; they liquefy, like gums, wax and so on. Thus the contrary effects of heat (some things being hardened by it, others liquefied) are easily explained; in the former case the spirit is released, in the latter it is stimulated but kept in. The latter is the action of the heat and spirit themselves; the former is the action of the tangible parts, the release of the spirit being merely the occasion.

But when spirit is neither wholly kept in nor wholly released, but only struggles and strives[71] within its bounds, and has possession of tangible parts which are obedient and tractable, so that where spirit leads, they immediately follow, the result is the formation of an organic body, the development of limbs and other activities of life, in both vegetables and animals. These things are best conveyed to the senses by careful observations of the earliest beginnings and rudiments or attempts at life in the tiny creatures which are born from putrefaction: as ants' eggs, worms, flies, frogs after rain, etc. For life to occur there has to be gentle heat and a pliant body, so that the spirit may not break out in a hurry, nor be prevented by the resistance of the parts from folding and fashioning them like wax.

Again, a great many instances of conveyance bring before our eyes the most remarkable and most wide-ranging of the distinctions between spirits: isolated spirit, simply branching spirit, and spirit which is both branching and cellular; of which the first is the spirit of all inanimate bodies, the second that of vegetables, the third that of animals.

It is likewise obvious that the more subtle structures and figures of things are not perceived or touched, though the bodies are completely visible or tangible. And therefore in these cases too information gets through by

---

[71] 'struggles and strives' (Kitchin)

conveyance. But the most radical and primary difference between structures depends on the greater or lesser amount of matter occupying the
same space or dimension. All other figures (which are attributable to the
particular features of the parts contained in the same body, and their
relative places and positions) are secondary to this one.

Suppose the nature under investigation is expansion or contraction of
matter in bodies respectively: viz. how much matter fills what dimension
in individual things. For nothing is truer in nature than the double propo-
sition that 'nothing comes from nothing', nor 'is anything reduced to
nothing',[72] but a given quantity of matter or the total amount is constant,
and neither increases nor diminishes. It is no less true, that 'from a given
quantity of matter more or less is contained within the same spaces
and dimensions, in accordance with the differences between bodies'; for
example, there is more in water and less in air. Hence to claim that a given
volume of water can be turned into an equal volume of air is like saying that
something can be reduced to nothing; and on the other hand to claim that
a given volume of air can be turned into an equal volume of water is like
saying that something can come from nothing. And the notions of *density*
and *rarity*, which are loosely taken in various ways, should properly be
derived from the larger and smaller amount of matter. We should also take
up a third claim which is also quite certain, that the amount of matter
which we say is in this or that substance can be reduced (by comparison) to
numbers, to exact or nearly exact measurements. For example, it would not
be wrong to say that in a given quantity of gold there is such an accumulation of matter that spirit of wine needs twenty-one times as much space as
the gold fills, to equal that quantity of matter.

Now the accumulation of matter and its measures are conveyed to sense
by weight. For weight corresponds to the amount of matter, so far as the
tangible parts of it are concerned, and spirit, and its material quantity, do
not come into the computation as weight; for it lightens the weight rather
than increases it. We have made a quite accurate table of this thing, in which
we have put down the weights and spaces of each metal, of the principal
stones, woods, liquids, oils, and most other bodies, both natural and artificial: a thing of many uses,[73] both to provide illuminating information and
to give a guide for operation; and which reveals much that is quite contrary
to expectation. It is also valuable in demonstrating that the full range of

[72] Bacon is here paraphrasing well-known scholastic axioms.
[73] *Polychrestam*: see II.50.

tangible bodies known to us (meaning compact bodies, not highly porous bodies, hollow and largely filled with air) does not exceed the ratio of 21:1; so limited is nature, or at least that part of it which is most relevant to our experience.

We have also thought it worth our while to see whether we could perhaps determine the ratio of intangible bodies to tangible bodies. We attempted this with the following contrivance. We took a small glass bottle, which could hold perhaps one ounce (we used a small vessel so that the consequent evaporation could be achieved with less heat). We filled this bottle with spirit of wine almost to the brim; selecting spirit of wine because we observed by means of an earlier table[74] that it is the rarest of the tangible bodies (which are continuous, not porous), and contains the least matter for its dimensions. Then we accurately noted the weight of the liquid[75] with the bottle itself. Next we took a bladder which would hold about two pints. We expelled all the air from it, so far as possible, to the point that the sides of the bladder were touching each other; we had also previously smeared the bladder with a grease, rubbing it gently in so that it would be more effectively closed, its porosity, if there was any, being sealed by the oil. We tied this bladder tightly around the mouth of the bottle, with its mouth inside the mouth of the bladder, lightly waxing the thread so that it would stick better and bind more tightly. Finally we placed the bottle above burning coals in a brazier. Very soon a steam or breath of spirit of wine, expanded by heat and turned into gaseous form, gradually inflated the bladder, and stretched the whole thing in every direction like a sail. As soon as this happened, we removed the glass from the fire, and placed it on a rug so that it would not be cracked by the cold; we also immediately made a hole in the top of the bladder, so that when the heat ceased, the steam would not return to liquid form and run down and spoil the measurement. Then we lifted up the bladder itself and again took the weight of the spirit of wine which remained. Then we calculated how much had been used up as steam or gas; and making a comparison as to how much place or space that substance had filled in the bottle when it was spirit of wine, and then how much space it filled after it had become gas in the bladder, we calculated the ratio; and it was absolutely clear that the substance thus converted and changed had achieved a hundredfold expansion over its previous state.

Similarly, suppose the nature under investigation is heat or cold in

[74] This table is given at *Historia densi et rari* (Ellis and Spedding, 11.245–6), according to Fowler.
[75] Translating *aquae* (water), but presumably 'spirit of wine' is intended.

degrees too weak to be perceived by the senses. These are conveyed to sense by a thermometer, such as we described above. For heat and cold are not perceptible to touch in themselves, but heat expands air and cold contracts it. And in turn it is not the expansion or contraction of the air which is perceptible to sight, but expanded air forces the water down, contracted air raises it; conveyance to the sight occurs only at that point, not before and not in any other way.

Similarly, suppose the nature under investigation is mixture of substances: i.e. what watery substance they contain, what oily substance, what spirit, ash, salt and so on; or (to take a case) what milk contains of butter, curds, whey and so on. These are conveyed to the sense by skilfully contrived separations, so far as their tangible elements are concerned. The nature of the spirit in them is not directly perceived, but detected in the various movements and tendencies which the tangible substances exhibit in the very act and process of their separation, and also in the bitterness, bite and different colours, smells and tastes of the same substances after the separation. And in this task men have surely made vigorous efforts, by distillations and contrived separations, but with no more success than in their usual manner of experimentation: groping methods, blind ways, more effort than intelligence, and (what is worst) no imitation or emulation of nature, but destruction (by high heat or excessively strong forces) of all the more subtle structure in which the hidden powers of things and their agreements chiefly lie. And the other caution we have given elsewhere[76] never troubles their thoughts or observations in such separations, that in the violent operations they perform on bodies, whether by fire or in other ways, many qualities are caused by the fire itself and by the substances used to make the analysis which were not in the compound before. Hence amazing errors. For instance, not all the steam which is released from water by fire was either steam or air previously in the substance of the water, but was formed for the most part by the expansion of the water by the heat of the fire.

Similarly, in general, this is the place to bring in all the refined ways of testing substances, whether natural or artificial, by which the true is distinguished from the adulterated, good quality from poor; for they convey the non-sensible to the sensible. They should therefore be sought everywhere with care and industry.

[76] II.7.

(5) As for the fifth way in which things hide themselves, it is obvious that the action of the senses occurs as motion, and motion occurs in time. If the motion of a body is either so slow or so fast that it is too slow or too fast to suit the speed at which the action of the senses takes place, the object is not perceived at all; as in the motion of the hand of a clock, or the movement of a bullet. The movement which is not seen because it is too slow is easily, and commonly, conveyed to the senses by the sum of its motions; motions which are too fast have not been able to be measured properly up to our time; however, the investigation of nature requires us to do this in some cases.

(6) The sixth case, where the sense is stymied by the object's power, achieves conveyance either by moving the object further away from the senses, or by dulling its effect by putting some barrier in front of it such as would weaken it without destroying it; or by admitting and receiving a reflection of it when the force of a direct strike would be too strong, like the reflection of the sun in a bowl of water.

(7) The seventh case in which an object fails to show up is the case where the sense is so oppressed by an object that there is no opportunity for any other object to make itself felt; this is more or less confined to smells and odours, and is not very relevant to the discussion. So much for the ways in which the non-sensible is conveyed to the sensible.

Sometimes a conveyance is made not to a man's sense, but to the sense of some other animal, which in some cases surpasses the human senses. For example, the conveyance of some smells to the sense of a dog; and of the light which exists latently in air which is not illuminated from outside itself, to the sense of a cat, an owl and other animals which see at night. Telesius rightly noted that there is indeed a kind of original light in the air itself, though weak and feeble and for the most part no use to the eyes of men or of most animals, because the animals whose senses are adapted to this kind of light see at night; and it is scarcely credible that this happens without light or by an internal light.

In any case, note that we are dealing here only with the defects of the senses and their remedies. For the errors of the senses should be assigned to the actual investigations of sense and the sensible; with the exception of the *great* error of the senses, that they set the outlines of things by the pattern of man, not of the universe;[77] which can only be corrected by universal reason and a universal philosophy.

[77] Cf. I.41, I.59.

XLI

In eighteenth place among the privileged instances we shall set *instances of the road* ,[78] which we have also chosen to call *travelling instances* and *jointed instances*. They are instances which indicate discretely continuous motions in nature. This kind of instance rather avoids observation than sense. Men are wonderfully inattentive here. In fact they observe nature only in a desultory and casual manner and after bodies are finished and complete, and not in its working. If you wanted to see the skills of a craftsman and observe his work, you would not only want to view the unformed materials of his craft, but also be there when he was at work himself and shaping his product. One must do something like that with regard to nature. For example, anyone investigating the growth of plants must observe it from the sowing of the seed on (this can easily be done by taking up, more or less every day, seeds which have been in the ground for two days, three days, four days and so on, and by carefully studying them); he must observe how and when the seed begins to bulge, swell and be filled with spirit (so to speak); and then how it begins to break the husk and put out shoots, at the same time pushing its way upwards a little bit unless the soil is very heavy; and how it also puts out shoots, some downwards for the roots, and some upwards as stems, and sometimes creeps sideways if it can find open and easier soil in that direction; and there are several other things he should observe. One should do the same for the hatching of eggs, where the process of life beginning and taking shape easily offers itself to view, and what and which parts come from the yolk, and what comes from the white of the egg, and so on. Animals from putrefaction offer a similar method. It would be inhuman to make such investigations of well-formed animals ready for birth by cutting the foetuses out of the womb, except for accidental abortions and in hunting and so on. And therefore one should keep a kind of round-the-clock watch on nature, since it offers itself for inspection better at night than in the day. For these observations might be regarded as nocturnal, because the lamp is so small yet everlit.

The same thing should be tried in the case of inanimate things, as we did in investigating the expansion of liquids through fire.[79] For there is one mode of expansion in water, another in wine, another in vinegar, another in the juice of grapes, and a very different one in milk, oil and so on. This

[78] *instantiae viae*
[79] II.40.

would be easy to see by boiling them in a glass vessel on a slow fire, where everything could be clearly seen. We are simply skimming this subject; we will discuss it more precisely and at greater length when we reach the discovery of the *latent process* of things. For we must always bear in mind that we are not dealing with the things themselves here, but only giving examples.

XLII

In the nineteenth place among the *privileged instances* we shall put *instances of supplement* or *substitution*,[80] which we have also chosen to call *instances of last resort*. They are the instances which supply information when the senses draw a total blank, and therefore we resort to them when we have been unable to get proper instances. This substitution happens in two ways; either by degrees or through analogies. For example: there is no medium found which completely stops the operation of a magnet in moving iron; you cannot do it by putting gold or silver between them, or stone or glass, or wood, water, oil, cloth or fibrous materials, or air or flame, and so on. And yet some medium might perhaps be found, by careful testing, which could dull its power more than anything else, relatively and in some degree: for instance, one might discover that a magnet does not draw iron through such a thickness of gold as it does through an equal distance of air; or through so much heated silver as cold silver; and so on in like cases. We have not made experiments with these; we simply include them as examples. Similarly, no body is found in human experience which does not take heat when brought near a fire. But air takes heat much more quickly than stone. Such is the substitution by degrees.

Substitution by analogy is certainly useful but less sure, and therefore must be used with some discretion. It occurs when a non-sensible thing is brought before the senses, not by sensible activity on the part of the insensible substance itself, but by observation of a related sensible body; for example, if in an investigation of a mixture of spirits, which are non-visible bodies, some kind of relationship is apparent between the bodies and their fuel or nourishment. The fuel of flames seems to be oil and fatty substances; of air, water and watery substances; for flames intensify over oil fumes, and air feeds on water vapour. We should therefore study the

[80] *instantiae supplementi, sive substitutionis*

mixture of water and oil, which is accessible to the senses, because a mixture of air and fire escapes the senses. But oil and water mix with each other very imperfectly when combined or stirred, but in grass, blood and the parts of animals the same things mix fully and smoothly. And therefore something similar could be the case in a mixture of fiery and airy parts in spirits: things which do not easily hold a mixture when simply poured together seem to mix in the spirits of plants and animals, especially as every living spirit takes in both kinds of moisture, the watery and the fatty, as fuel.

Similarly, we may be investigating not a relatively complete mixture of spirits, but a mere combination, that is whether spirits are readily incorporated into each other, or whether (for example) there are some winds and vapours or other spirituous bodies which do not mix with ordinary air, but only hang and float in it, in globules and drops, and are smashed and shattered by air, not welcomed and incorporated. This is not perceptible to the senses in ordinary air and other spirituous bodies, because they are too subtle; however, a kind of image of the thing, so far as it happens, may be drawn from the liquids quicksilver, oil and water; and also from the breakup of air when it is dispersed and rises through water in small particles; also in thick smoke; and finally in disturbed dust floating in the air; in all of which there is no incorporation. The presentation we made on this subject is not bad, provided a careful inquiry is first made whether there can be such heterogeneity among spirits as is found among liquids; for in that case substituting these images by analogy will not be inappropriate.

We have said that information may be got from these *supplemental instances* as a last resort when direct instances are lacking. But we also want to make it clear that they are also very useful when direct instances are available, for confirming the information along with the direct instances. But we will explain them more exactly when the regular course of our discussion brings us to a treatment of *the supports of induction*.

XLIII

In the twentieth place among the *privileged instances* we shall put *cleaving instances*,[81] which we have also chosen to call *plucking instances*, but for a

---

[81] *instantiae persecantes*

different reason. We call them *plucking* instances because they pluck at the mind, *cleaving* because they cleave nature apart; so too we sometimes call them *instances of Democritus*.[82] They are instances which remind the intellect of the exquisite subtlety of nature, so that it is excited and prompted to give nature the attention, observation and scrutiny it deserves. For example: that a little drop of ink can make so many letters or lines; that a piece of silver, gilded only on the outside, makes such a length of gilt wire; that a tiny little worm such as is found in the skin contains spirit and a distinct structure of parts; that a little saffron stains and colours a whole barrel of water; that a little civet or perfume fills a far greater quantity of air with its smell; that a little fumigant makes such a cloud of smoke; that the precise discriminations of sound articulated by voices speaking are somehow carried through the air, and penetrate even the passages and pores of wood and water (though in diminished form), and indeed are echoed back so swiftly and distinctly; that light and colour rapidly permeate, even at a great distance, the solid substance of glass and water, and fill it with such an exquisite variety of images, and are also refracted and reflected; that a magnet is effective through every kind of body, even the most solid. More wonderfully still, in all these things, the action of one of them in a neutral medium such as air does not greatly impede another; so that at the same time the spaces of the air carry so many images of visible things, so many resonations of voices speaking, and so many distinct perfumes, as of the violet and the rose; also heat and cold and magnetic powers; all (I repeat) at the same time, and without obstructing each other, as if each one had its own ways and private paths apart, and none touched or crossed any other.

But there is a useful appendix to add to *cleaving instances*, and that is what we call the *limits of dissection*; that in the things we have spoken of, an action does not impede or disturb a different kind of action, but may subdue and extinguish another action of the same kind: as the light of the sun does to the light of a candle; the sound of a bomb to the voice; or as a stronger odour overcomes a more delicate one; and intenser heat a lesser heat; and sheets of iron put between a magnet and another piece of iron impedes the operation of the magnet. The proper place to deal with these will also be among the *supports of induction*.

---

[82] Cf. 1.51.

XLIV

We have now dealt with the instances which aid the senses, which are of particular value for the informative part of our enterprise. For information starts from sense. But the enterprise as a whole ends in works; this is the end of the thing, as information is the beginning. Instances therefore will follow which are of particular value for the applied part. They are of two kinds, and there are seven of them. We have chosen to give all of them the general name of *practical instances*. There are two faults in the *operative part*, and the same number of leading instances in general. For either an operation fails or causes too much labour. The major reason why an operation fails (especially if there has been a careful search for the natures) is that the strengths and actions of the bodies have not been properly determined and measured. The strengths and actions of bodies are described and measured by dimensions of space, or by moments of time, or by units of quantity, or by the dominance of a power;[83] if these four factors are not honestly and carefully measured, it will perhaps make a pretty, speculative science, but it will be empty of results. Likewise we give the four instances which correspond to these the single name of *mathematical instances*, and *instances of measurement*.

Practice becomes laborious, either because of a clutter of useless subjects or an excessive number of instruments, or because of the mass of matter and substances which happen to be required for some task. Therefore we should value instances which either direct the operative function to things of most value for men, or which keep down the number of instruments, or which economise on material or equipment. The three instances which belong here we call by the one name of *propitious* or *benevolent instances*. And therefore we shall now speak of these seven instances one by one; and with them we shall close this part of the privileged or leading instances.

XLV

In the twenty-first place among the *privileged instances* we shall put *instances of the rod* or *of the ruler*,[84] which we have also chosen to call *instances of range* or *furthest limit*. For the powers and motions of things work and have effect at distances which are not indefinite and a matter of chance, but are fixed

---

[83] These subjects are dealt with in 11.45, 46, 47 and 48, respectively.
[84] *instantiae virgae, sive radii*

and definite. It is of great practical value to observe and note them in each nature we seek, not only to avoid error in practice but also to improve practice and extend its power. For there is sometimes the opportunity to enlarge powers and draw distances closer, as in the case of magnifying glasses.[85]

Most powers operate and have an effect on other things only by open contact, as in the case of the collision of bodies, when one body does not move the other without the striking body actually touching the body struck. Also medicines applied from the exterior, like ointments and plasters, only exercise their powers by bodily contact. The objects of the senses of touch and taste affect them only when they are close to the organs.

There are other powers which work at a distance, albeit a very small distance. Few have yet been noted, though there are more than men imagine. For instance (to take examples from common objects), amber or jet attracts straws, a bubble bursts another bubble when it gets close to it, certain purgatives draw down rheum, and so on. And the magnetic power which draws iron to a magnet or draws magnets to each other works within a certain range of power, though a small one; whereas if there is a magnetic power proceeding from the earth itself (evidently just below the surface) on to an iron needle and affecting its polarity, the effect would be working at a great distance.

Again, if there is any magnetic force operating by agreement between the globe of the earth and heavy things, or between the globe of the moon and the waters of the sea (which seems very likely in the high and low tides twice a month), or between the starry heaven and the planets, by which they are called up and raised to their apogees; all these things would be operating at very great distances. There are also some cases of fires starting or breaking out in some materials at quite large distances, as they report of the naphtha at Babylon. For heat travels vast distances, and so does cold: the inhabitants of Canada feel from far away the cold given off by the masses or bulks of ice which break off and float through the northern ocean and move down through the Atlantic towards their shores. Odours too are effective at remarkable distances (though there seems to be also some physical emission in their case), as noted by sailors along the coasts of Florida or some parts of Spain, where there are whole forests of lemon trees, orange trees and such fragrant plants, or stands of rosemary,

[85] See II.39.

marjoram and so on. Finally, light radiation and sound impressions work at considerable distances.

But whether the distance at which they work is great or small, all these things certainly work at distances which are fixed and known to nature, so that there is a kind of *No Further* which is in proportion to the mass or quantity of the bodies; or to the vigour or weakness of their powers; or to the assistance or resistance of the surrounding medium; all of which should come into the calculation and be noted. Furthermore we should also note the measures of the so-called violent motions, as of missiles, cannon, wheels and so on, since they clearly have their fixed limits also.

There are also certain motions and powers which are contrary to those which operate on contact and not at a distance: i.e. those which operate at a distance and not by contact, and those which operate weakly at a short distance and more strongly at a greater distance. Vision for instance is not communicated well by contact but needs a medium and a distance. But I do remember having heard a story from someone who deserved to be believed, that when he had the cataracts on his eyes treated (the treatment was to insert a little silver needle under the first membrane of the eye, to remove the film of the cataract and push it into the corner of the eye) he saw the needle moving over the actual pupil very clearly. However true this may be, it is obvious that larger objects are not well or clearly seen except at the point of a cone, where the rays from the object meet at some distance. Furthermore, in old people the eye sees better when the object is placed a little further away than nearer. And in the case of missiles it is certain that the impact is not so great at a very short distance as it is a little way away. These and similar things are what we should note in measuring motions with regard to distance.

There is also another kind of spatial measurement of motion which should not be ignored. This belongs not to linear motions but to spherical motions, i.e. to the expansion of bodies into a larger sphere, or their contraction into a smaller. Among these measurements of motion we must ask how much compression or expansion bodies easily and freely allow (in accordance with their natures), and at what point they begin to resist, so that at the extreme they bear it *No Further*; as when an inflated bladder is compressed, it tolerates some compression of the air, but after a point it can bear it no longer, and the bladder bursts.

We have tested this more precisely with a subtle experiment. We took a small metal bell, quite thin and light, like a saltcellar, and sank it in a basin of water, so that it took with it to the bottom of the basin the air held in its cavity. We had first placed a little ball on the bottom of the basin on which to set the bell. The result was that if the ball was quite small (in relation to the cavity) the air retreated into a smaller area, and was simply compressed and not expelled. But if it was too large for the air to give way freely, then the air could not tolerate the greater pressure but partially lifted the little bell and came up in bubbles.

To find out what extension (no less than compression) air would permit, we did the following experiment. We took a glass egg with a little hole at one end of it. We drew the air through the hole by strong suction, and immediately blocked the hole with a finger, and immersed the egg in water, and then removed the finger. The air was put under pressure by the tension induced by the suction, and swollen beyond its own nature, and in striving to draw back and contract (so that if the egg had not been immersed in water, it would have drawn air itself in with a whistle) it drew in a sufficient quantity of water for the air to recover its former sphere or dimension.

It is also certain that the more tenuous bodies (like air) permit some noticeable contraction, as has been said; but tangible substances (like water) permit compression with much more reluctance and to a lesser degree. How much it permits was the object of this experiment.

We had a hollow globe made of lead, to hold about two vintners' pints, and with sides thick enough to bear considerable force. We put water into it through a hole made in it; and after the globe filled with water, we sealed the hole with liquid lead, so that it became a completely solid globe. Then we flattened the globe on two opposite sides with a heavy hammer; as a result of which the water had to be contracted into a smaller space, since a sphere is the most capacious of shapes. Then when hammering could do no more, as the water was resisting retreat, we used a mill or screw press; so that at last the water could tolerate no more pressure, and distilled through the solid surface of the lead (like a light dew). Then we calculated how much the space had been reduced by the compression; and inferred that the water (but only when submitted to so much force) had suffered that amount of compression.

More solid, dry or compact materials, such as stone and wood, and also metals, tolerate much less compression or extension; in fact, it is almost imperceptible; they free themselves by breaking, by moving, or other manoeuvres, as appears in bending wood or metal, in clocks that move by the winding of a spring, in projectiles, in hammering and in countless other motions. All these things with their measurements should be noted and tested in the hunt for nature, either in their own exact form, or by estimate, or by comparison, as opportunity arises.

XLVI

In the twenty-second place among *privileged instances* we shall put *running instances*,[86] which I have also chosen to call *instances of water*, taking the term from the water clocks of the ancients, which were filled with water rather than sand. They measure nature by moments of time, as *instances of the rod* measure them by units of space. For every natural motion or action passes in time, one more swiftly, another more slowly, yet all at moments which are certain and known to nature. Even actions which seem to happen instantly and in the twinkling of an eye (as we say) are found to involve more or less of time.

First then we see that the revolutions of the heavenly bodies take place at calculable intervals, and so do the ebb and flow of the sea. The movement of heavy things towards the earth and of light things towards the circuit of the sky occupies certain moments determined by the body which is moving and the medium through which it moves. The paths of ships, the movements of animals, the courses of missiles, all likewise occupy lengths of time which may be reckoned (in round terms). As far as heat goes, we see boys wash their hands in flames in winter and yet not be burned, and jugglers make smooth, agile movements to turn jars full of wine or water upside down and right way up again without spilling a drop; and much of the same kind. Equally, the compression, expansion and eruption of bodies happen quickly or slowly depending on the type of body and the motion, but all take a certain amount of time. Further, in the explosion of several cannon at the same time, which may sometimes be heard up to thirty miles away, those who are near the spot where the sound occurred hear it before those who are further away. And in sight (whose action is very swift) it is also clear that certain moments of time are needed for it to act; that is proved by things which are too swift to be seen, as by the path of a bullet from a gun. For the bullet flies past too fast for an impression of its species to reach the eye.

This and other things have sometimes given us a quite fantastic doubt

[86] *instantiae curriculi*

as to whether we see the face of the serene and starry heaven at the same time as it truly exists, or some time later; and whether (so far as the sight of heavenly bodies is concerned) there is both a true time and a perceived time, as in the case of parallaxes, where astronomers have remarked that there is a true place and a perceived place. So incredible did it seem to us that the species or rays of celestial bodies could pass instantly to the sight through such immense distances of miles, rather than travel in some noticeable length of time. But that doubt (as to some appreciable interval of time between the true time and the perceived time) later completely vanished as we reflected on the infinite loss and decrease of quantity, so far as appearance goes, between the true body of the star and the image perceived, which is caused by the distance; and noted also at what a distance (a minimum of sixty miles) bodies which are merely white are instantly seen here on earth, since there is no doubt that the light of the heavenly bodies many times exceeds, in vigour of radiation, not only the brightness of white, but also the light of every flame known to us here on earth. Also the immense velocity of the body itself as seen in its daily motion (which has so amazed even serious thinkers that they would prefer to believe in the motion of the earth) makes the motion of ray emission (marvellously fast, as we said) more credible. But the most convincing point is that if some noticeable interval of time were put between the reality and the sighting, the species would frequently be intercepted and muddled by clouds arising in the meantime and similar disturbances in the medium. So much for simple measurements of time.

But we should investigate the measurement of motions and actions not only in themselves but, much more, comparatively; this is supremely useful and for many purposes. We see that the flash of a gun firing is seen more quickly than the sound is heard, though the ball must have hit the air before the flame behind it could get out; and this must happen because of the swifter passage of the motion of light than of the motion of sound. We see also that visible images are taken up by sight more quickly than they are discarded; which is why violin strings plucked with a finger are doubled or tripled in appearance, because a new image is received before the old one is discarded; hence rotating rings look like a globe, and a burning torch carried swiftly at night seems to have a tail. On this foundation of the unequal speed of motions Galileo built his conception of the ebb and flow of the sea: the earth rotates more swiftly, and water less swiftly, and so the waters heap up and then in turn fall down again, as is displayed in a vase of water moved rapidly. He achieved this fiction by granting himself the ungrantable (namely that the earth moves), and without being well informed on the six-hourly motion of the ocean.

We are discussing the comparative measure of motions, both in itself and in its eminent utility (of which we spoke just now). A remarkable example occurs in mines laid beneath the earth and filled with gunpowder; in which a tiny quantity of gunpowder undermines and throws into the air immense masses of earth, buildings and so on. The cause of it is certainly that the expansive motion of the powder, which is the impelling force, is many times swifter than the motion of gravity by which some resistance could be made; so that the first movement has finished its task before the contrary motion begins; hence at the beginning there is an absence of resistance. This is also the reason why in every missile, the blow, which is not so strong as it is sharp and swift, has a very high projectile power. And the only reason why a small quantity of animal spirits in animals, especially in such large-bodied animals as whales or elephants, could steer and control such an immense bodily mass is that spirit motion is very fast, and physical motion is slow to exert resistance.

Finally, this is one of the main foundations of magic experiments, which we shall discuss later, namely when a small amount of matter masters a much larger amount and regulates it; this, I say, occurs if the one forestalls the other with its speed of motion before the other gets moving.

One more thing: the *before* and *after* should be noted in every natural action. For instance, in an infusion of rhubarb, the purgative force comes out first, and then the astringent; we have seen something similar in an infusion of violets in vinegar, where the sweet, delicate scent of the flower is noticed first and then the more earthy part of the flower, which overwhelms the scent. Hence if violets are infused for a whole day, the scent is much weaker to notice; but if they are infused only for a quarter of an hour, and then taken out, and (because there is little scented spirit within a violet) new, fresh violets are infused, for a quarter of an hour each, six times; then at last the infusion is made so strong that, though the violets are only in it (one bunch at a time) for a total period of an hour and a half, still a very pleasing scent, as good as the real violet's, lasts for a whole year. However, one should note that the scent does not reach its full strength till a month after the infusion. And in a distillation of aromatic spices soaked in spirit of wine, it is evident that at first a useless, watery fluid develops, then water with more spirit of wine in it, and only after that, water with more scent. A great many such things are found in distillations and deserve to be noticed. But this will be enough to give examples.

XLVII

In the twenty-third place among the *privileged instances* we shall put *instances of quantity*,[87] which we have also chosen to call *doses of nature* (borrowing a term from medicine). These are the instances which measure powers by the quantities of bodies, and indicate what *quantity of a body* results in a certain *amount of power*. And first there are some powers which exist only in *cosmic quantity*, i.e. in a *quantity* which is consistent with the shape and structure of the universe. For the earth stands, its parts fall. Water in the sea ebbs and flows, but not water in the rivers, except as the sea flows into them. Again, the effect of almost all the powers depends on whether there is a *lot* or a *little* of the substance. Large bodies of water are not easily polluted, small ones quickly. Wine and beer mature and become drinkable much more quickly in small leather bags than in large casks. If a herb is placed in a larger amount of liquid, the herb is infused, and the liquor is not absorbed; if in a smaller amount, there is no infusion and the liquid is absorbed. In its effect on the human body, a bath is one thing, a light sprinkling another. Light dews in the air never fall; they are dissipated and incorporated into the air. Breathing on a jewel, one may see that little bit of moisture instantly dissolved, like a small cloud blown away by the wind. Also a fragment of magnet does not attract so much iron as a whole magnet. There are also powers in which a small *quantity* has the greater effect, as in making holes a sharp point pierces more quickly than a blunt one, a pointed diamond engraves on glass, and so on.

But we should not linger too long on this; we have also to investigate the *proportion of quantity* of body to amount of power. It would be natural to think that the proportions of quantity equal the proportions of power; so that if a lead ball weighing an ounce took a certain amount of time to fall, a ball of two ounces should fall twice as quickly, which is quite wrong. Nor do the same proportions hold in all the different kinds of powers, but very different ones. And therefore we must look for these measures in the things themselves, not on the basis of likelihood or conjecture.

And last, in every inquisition of nature we must note *how much* of a body

---

[87] *instantiae quanti*

is needed for each particular effect; and keep slipping in warnings about *too much* and *too little*.

XLVIII

Twenty-fourth among the *privileged instances* we shall put *instances of struggle*,[88] which we have also chosen to call *instances of dominance*. They point to alternating dominance and submission of powers, and indicate which of them is stronger and prevails, and which is weaker and succumbs. For the motions and exertions of bodies are compounded, decompounded and combined with each other no less than the bodies themselves. We shall therefore first give an account of the main kinds of motion or active power, in order to offer a clearer comparison of their respective strengths, and on that basis to exhibit and demarcate *instances of struggle and dominance.*

(1) Let the first motion be the motion of the *indestructibility* of matter, which exists in every little part of it, by which it utterly refuses to be annihilated; so that no fire, no weight, no pressure or violence, nor age nor length of time, can reduce even the smallest particle of matter to nothing; but it always is something and occupies some space, and (under any necessity) either frees itself by changing its shape or place, or (if it has no chance to do so) it remains as it is, and never gets to the point of being nothing or nowhere. This motion the School (which nearly always names and defines things rather by their effects and negative consequences than by their inner causes) either denotes by the axiom that 'two bodies cannot be in one place; or calls it the motion not to permit penetration of dimensions'. We need not give examples of this motion: it is in every substance.

(2) Let the second motion be the motion we call *bonding*; by which bodies refuse to be torn in any part of themselves from contact with another body, as if they enjoy mutual bonding and contact. This motion the School calls the motion *to avoid a vacuum*; as when water is drawn up by suction or through a syringe, or flesh by cupping-glasses; or when water stands and fails to run out through a hole in a water pot unless the mouth of the jar is opened to let in air; and innumerable such things.

(3) Let the third motion be the motion of *liberty* (as we call it); by which bodies strive to free themselves from unnatural pressure or tension and to restore themselves to the dimensions that suit their body. There are count-

---

[88] *instantiae luctae*

less examples of this motion: as (for liberation from pressure) the example of water in swimming, of the air in flying; of water in rowing, of air in blasts of wind; of the spring in clocks. The motion of compressed air shows itself rather neatly in children's popguns, when they hollow out a length of alder or some similar thing, and stop up both ends with a piece of some pulpy root, or something of that kind; then with a ramrod they stuff a root or some such wad in at one end; the root at the other end is forced out and ejected with an audible noise before being touched by the root or wad at the nearer end, or with the ramrod. As for escape from tension, this motion shows itself in the air which remains in glass eggs after extraction, and in strings, leather and cloth which resume their shape when the stretching is finished, unless it has lasted long enough to become permanent, etc. This motion the School refers to under the name of motion from *the form of the element*: ignorantly enough, since this motion has to do not only with air, water or fire, but with the whole spectrum of solid bodies, as wood, iron, lead, cloth, parchment etc., in which each body has a limit of its own characteristic dimension, and is with difficulty forced out of it to any noticeable extent. But as the motion of liberation is the most evident of them, and has an infinite number of forms, it would be advisable to make some good, clear distinctions. For some men carelessly confuse this motion with the dual motion of *indestructibility* and *bonding*; i.e. they confuse liberation from pressure with the motion of *indestructibility*, and liberation from tension with the motion of *bonding*; as if bodies under compression submitted or expanded to avoid *penetration of dimensions*, and bodies under tension recoiled and contracted to avoid a *vacuum*. However, if compressed air attempted to contract to the density of water, or wood to the density of stone, there would be no need of *penetration of dimensions*, and yet it would be a much greater compression of them than they in any way allow. In the same way, if water tried to expand and achieve the rarity if air, or stone the rarity of wood, there would be no need of a *vacuum*, yet there would be a much greater extension of them than they in any way allow. Therefore this is not a matter of *penetration of dimensions* and *vacuum*, except in the last stages of condensation and rarefaction. These motions stay and stop long before those stages are reached, and are simply the efforts of bodies to keep their own consistencies (or, if they[89] prefer, their own forms) and not to lose them suddenly unless they are altered by gentle means and with their own

[89] 'They' refers to the scholastics.

consent. But it is much more essential (because of the many consequences) to impress upon men that violent motion (which we call *mechanical*, and Democritus, who in explaining his first motions is to be ranked even below mediocre philosophers, called the motion of the *blow*) is simply a motion of liberty, i.e. from compression to relaxation. For whether it is a simple impulsion or a flight through the air, there is no displacement or spatial movement until parts of the body suffer unnaturally by compression by the impellent. Then as each part pushes another one after the other, the whole body moves; not only going forward but also rotating at the same time; so that in this manner the parts too may be able to escape or share the burden more equally. Enough on this motion.

(4) Let the fourth motion be the motion to which we have given the name of *matter*.[90] This motion is in some way the converse of the motion of liberty just discussed. For in the motion of liberty bodies abhor, reject and avoid a new dimension or a new sphere or a new expansion or contraction (these various words mean the same thing), and strive with all their might to rebound and regain their old consistency. By contrast, in this motion of matter bodies seek a new sphere or dimension, and seek it freely, eagerly and sometimes with the most powerful effort (as in the case of gunpowder). The instruments of this motion, not the only ones to be sure but the most powerful or at least the most frequent, are heat and cold. For example: if air is expanded by tension (as by suction in glass eggs), it makes great efforts to restore itself. But if heat is applied, it has on the contrary a positive desire to expand, it covets[91] a new sphere, it passes and crosses over to it freely as into a new form (as people call it); and after some expansion it no longer cares to return unless provoked to do so by the application of cold; which is not a return but a second transformation. In the same manner too, if water is constricted by compression it rebounds, and tries to be what it was, namely more diffuse. But if intense, continual cold sets in, it changes itself freely and voluntarily into the dense matter of ice; and if the cold continues without a break, uninterrupted by a warm spell (as happens in caves and caverns of any depth), it turns to crystal, or a similar substance, and never reverts.

(5) Let the fifth motion be the motion of *cohesion*. We do not mean simple primary cohesion with another body (that is the motion of *bonding*),

---

[90] 'Matter' translates *hyle*.

[91] Reading *concupiscit* with the second edition (Amsterdam, 1660) for the *concupiscet* of the first edition (1620).

but self-cohesion in a single body. It is quite certain that bodies abhor the dissolution of their cohesion, some more, some less, but all to some degree. In hard bodies (like steel or glass) the resistance to dissolution is very strong and vigorous, but also in liquids, where this motion seems to be lacking or at least very weak, still it is not found to be completely absent, but is clearly in them in a very low degree, and reveals itself in many experiences, for instance in bubbles, in the roundness of drops, in fine threads of water running from a roof, in the stickiness of viscous bodies, and so on. This tendency shows itself best if one attempts to break something up too small. For the pestle can do no more in a mortar after the substance has been crushed to a certain point; water does not get into the finest cracks; and even air, despite the subtlety of its own body, does not instantly pass through the pores of reasonably solid containers, but only by long seepage.

(6) Let the sixth motion be what we call the motion for *gain*, or the motion of *want*. This is the motion by which, when bodies are involved with things which are completely different in kind from themselves and almost hostile to them, if they get the chance and the opportunity to avoid those unsympathetic bodies and to attach themselves to more congenial things (even though these congenial things have no close agreement with themselves), still they instantly embrace them, and prefer them as better. They seem to regard it as a gain (and hence the word we have chosen) as if they were in want of such bodies. For example: gold leaf, like other metals in leaf form, does not enjoy being surrounded by air. Hence if it can get hold of a thick, tangible body (finger, paper, whatever), it instantly attaches to it, and is not easily got off. Also paper or cloth and things of that kind are not happy with air entering them and mingling in their pores. And therefore they gladly soak up water or liquid, and exclude the air. Also lumps of sugar or sponges soaked in water or wine gradually and by degrees draw the water or wine up, even if they stick out a long way above the surface.

Hence we get the best rule for opening and dissolving bodies. Apart from corrosives and acids, which have their own ways of opening, if an appropriate body can be found which is more agreeable and akin to some solid body than the one with which it is forcibly united, the body instantly opens up and relaxes, and takes it into itself, rejecting and excluding the other. This *motion for gain* does not operate or have effect only on contact. For the electrical operation (on which Gilbert and others after him have spread such stories) is simply the appetite of a body excited by a light friction, which does not tolerate air and prefers whatever other tangible body it can find close by.

(7) Let the seventh motion be the motion of *major aggregation* (as we call it), by which a body is drawn towards a mass of bodies of similar nature: heavy substances to the globe of the earth, light things to the circuit of the sky. The School has marked this with the name of natural motion, for the trivial reason, apparently, that there was nothing visible from the outside to start that motion (and so it must, they thought, be innate and inborn in the things themselves); or perhaps because it never stops. And no wonder: because heaven and earth are always there; whereas by contrast the causes and origins of most of the other motions are sometimes present and sometimes absent. And therefore the School defined this motion as native and perpetual, and the rest as artificial, because it is not intermittent but instantly starts as soon as the others pause. But in truth this motion is quite weak and feeble, since (except where there is a large mass of a body) it gives way and succumbs to the other motions. And though this motion has so filled men's thoughts that it has thrown other motions into the shadow, still there is little that men know of it, but they are involved in many errors about it.

(8) Let the eighth motion be the motion of *minor aggregation*, by which the homogeneous parts in a body separate from heterogeneous parts and coalesce among themselves; by which also whole bodies join and embrace each other because their substance is similar, and sometimes drift together from a distance, are attracted and unite: as when cream gradually comes to the top of the milk, and the lees and dregs of wine sink to the bottom. These things are not caused so much by weight or lightness, so that some parts make for the top, while others tend to the bottom, but much more through the desire of homogeneous things to come together and combine with each other. This motion differs from the motion of *want* in two ways. One, that in the motion of *want* the major stimulus is that of an evil, contrary nature, but in this motion (provided there are no obstacles and bonds) the parts unite through friendship, even though there is no foreign nature to cause conflict; secondly, that the union is closer and more a matter of choice. For in the motion of *want*, bodies which are not closely related combine, merely avoiding foreign bodies, whereas in this motion there is a union of bodies which are bound by a fully kindred likeness, and are fused into one. This motion takes place in all compound bodies and would be easy to observe in each one if it were not bound and bridled by the other tendencies and necessities of bodies which disturb the union.

This motion is inhibited usually in three ways: by the sluggishness of bodies, by the constraint of a dominant body and by an external motion. As for sluggishness: it is certain that there is in tangible bodies a kind of sloth to a greater or lesser degree, and an aversion to spatial motion, so that unless stimulated, they would be content with their own state (whatever it may be) rather than go to the trouble to get into a better state. This kind of sluggishness is thrown off in three different ways: by heat, or by the superior power of a related body, or by a lively and powerful motion. And first for the assistance of heat: this is the reason why heat has been said to be what *separates heterogeneous things and unites homogeneous things.* Gilbert has rightly rejected this definition of the Peripatetics[92] with contempt. It is as if, he says, one were to claim and define man as a thing that sows wheat and plants vines; for it is a definition merely by effects, and particular effects at that. But the definition is yet more vulnerable, because those effects (whatever they may be) do not come from the property of heat except accidentally (for cold has the same effect, as we shall argue later); they come from the desire of homogeneous parts to combine, where heat merely helps to shake off the sluggishness, which was what had previously checked the desire. As for the help of a power offered by a related substance: there is a wonderful illustration of this in the armed magnet, which excites in iron the virtue of retaining iron because their substance is similar; the sluggishness of the iron is shaken off by the power of the magnet. As for the assistance of motion, this is best seen in wooden arrows which also have wooden tips; after the swift motion has shaken off the sluggishness of wood, they penetrate more deeply into other pieces of wood than if they had been armed with iron because the substance is the same. We also discussed these two experiments in the aphorism on *concealed instances.*[93]

The constraint on the motion of *minor aggregation* which is caused by restraint from a dominant substance is seen in the dissolution of blood and urine by cold. As long as those substances are filled by an active spirit, which is master of the whole and regulates and restrains their individual parts of every kind, for so long the different parts do not coalesce, because they have this check on them. But when the spirit has evaporated, or has been suffocated by the cold, then the parts are released from restraint and follow their natural desire to combine. This is why all bodies which contain a sharp spirit (like salts, and things of that kind) are not dissolved

[92] Aristotle and his school.
[93] II.25.

but endure, because of the permanent and durable restraint of the dominant and ruling spirit.

But constraint on the motion of *minor aggregation* caused by an external motion is best seen in the disturbance of bodies which prevents putrefaction. For all putrefaction is based on a combination of homogeneous parts; as a result of which the former nature (as they call it) is gradually corrupted and a new one generated. For the putrefaction which paves the way to the generation of new forms is generally preceded by the dissolution of the old forms, and is itself a combination to create homogeneity. If it is not impeded, a simple dissolution takes place; if obstructions of various kinds occur, putrefactions follow which are the beginnings of a new generation. However, if there is frequent disturbance from an external movement (which is our present concern), then the motion of combination (which is delicate and sensitive and needs protection from external movements) is disturbed and ceases. We see this happen in innumerable cases: as when daily stirring or discharge of water prevents putrefaction; winds prevent air pestilence; grain in granaries remains pure if it is turned and stirred; in fact, anything that is disturbed from outside does not easily rot inside.

It remains to deal with the combination of the parts of bodies which is the chief cause of hardening and drying out. For in a porous body (wood, bone, parchment etc.), after the spirit or the moisture which has turned into spirit has evaporated, then the denser parts contract and combine with greater force, and the result is hardening and drying out. This happens, we think, not so much by a motion of bonding to avoid a vacuum as by a motion of friendship and union.

As for combination at a distance, it is uncommon and rare, but occurs in more cases than is recognised. Here are some likenesses of it: when one bubble dissolves another; when medicines draw out humours because their substance is similar; when a string in one stringed instrument causes a string in another to make the same sound, and so on. We also think that this motion is vigorous in animal spirits, though this goes completely unrecognised. It is certainly evident in the magnet and in magnetised iron. However, when we speak of magnetic movements, we have to make a sharp distinction. For there are four powers or operations in a magnet, which should not be confused but kept distinct; though men's wonder and bewilderment have mixed them up. One is the union of magnet with magnet, or of iron with magnet, or of magnetised iron with iron. The second is its north–

south polarity and deviation from that. The third is its penetration through gold, glass, stone, everything. The fourth is the communication of power from the stone to iron and from iron to iron without communication of substance. Here we are speaking only of its first power, that of combination. Remarkable too is the combining motion of quicksilver and gold; so that gold attracts quicksilver, even when made up into ointments; and men who work among quicksilver vapours have a habit of holding a piece of gold in their mouths to collect the emissions from the quicksilver, which would otherwise attack their heads and bones; whence too the piece of gold soon turns white. So much for the motion of *minor aggregation*.

(9) Let the ninth motion be the *magnetic motion*; it is in general a motion of *minor aggregation*, but if it operates at great distances and on great masses of things, it deserves a separate investigation, especially if it does not start from contact as most motions do, nor continue the action till contact occurs, as all aggregating movements do, but only lifts bodies or makes them swell, and nothing further. For if the moon raises the waters, or makes moist things swell or expand; if the heaven of stars draws the planets towards their apogees; or the sun keeps the stars of Venus and Mercury at a certain distance from its body and no further, it does not seem appropriate to list these motions as major or minor aggregations. They seem to be intermediate and imperfect forms of aggregation, and should therefore form their own kind.

(10) Let the tenth motion be the *motion of avoidance*, a motion that is contrary to the motion of *minor aggregation*. In the *motion of avoidance* bodies flee out of antipathy and scatter hostile bodies, and separate themselves from them, and refuse to mingle with them. This motion may seem in some ways to be only a motion accidentally and consequentially, parasitic on the motion of *minor aggregation*, because homogeneous things cannot coalesce without excluding and getting rid of heterogeneous things. But it must be classified as a motion in itself and made a species, because in many things the desire for *avoidance* is seen to override the appetite for combination.

This desire is remarkably prominent in the case of animals' excretions, and no less so in objects offensive to some of the senses, especially smell and taste. For a fetid odour is so fiercely rejected by the sense of smell that it even induces, by agreement, a motion of expulsion in the mouth of the stomach; and a nasty, bitter taste is so fiercely rejected by the palate or the throat that by agreement it causes a shaking of the head and a shiver. This motion also takes place in other things. It is observed in some oppositions,[94] as in the middle region of the air, whose cold seems to be an exclusion of the nature of cold from the confines of the heavenly bodies; just as those great burning heats and fires that are found in subterranean regions are exclusions of the nature of heat from the interior of the earth. For heat and cold, in small quantities, mutually exterminate each other; but if they occur in larger masses, in full force so to speak, then indeed they struggle to exclude and eject each other from places. They also say that cinnamon and sweet-smelling things retain their smell longer when placed next to latrines and malodorous spots, because they refuse to come out and mix with the filthy odours. Certainly quicksilver is prevented from returning to its whole form, as it would otherwise do, by human saliva, or axle-grease,[95] or turpentine, and suchlike, which prevent its parts from uniting because of their lack of agreement with bodies of this kind. When surrounded by them, they withdraw; and thus their *avoidance* of the intervening substances is stronger than their desire to unite with the parts that are like them; this is what they call the *mortification* of quicksilver. The fact that oil does not mix with water is not only because of the difference of weight, but also because there is little agreement between them; as may be seen from spirit of wine, which is lighter than oil but mixes well with water. Most remarkable of all is the motion of *avoidance* in nitre, and such crude bodies, which have a horror of fire; as in gunpowder, quicksilver and also gold. However, iron's *avoidance* of one pole of the magnet is well remarked by Gilbert not to be *avoidance* in the proper sense, but conformity and acceptance of a more suitable position.

(11) Let the eleventh motion be the *motion of assimilation*, or of *self-multiplication*, or *simple generation*. By *simple generation* we do not mean the generation of whole bodies, as in plants and animals, but of similar bodies. By this motion similar bodies change other bodies which are akin to them, or at least sympathetic and prepared, into their own substance and nature: like flame, which multiplies on vapours and oily substances, and generates new flame; air, which multiplies over water and watery substances, and generates new air; vegetable and animal spirit, which multiplies over the more delicate parts of both watery and oily substances in food, and generates new spirit; the solid parts of plants and animals, like leaves, flowers, flesh, bone and so on, each of which assimilates and generates new

---

[94] *antiperistasis*; cf. II.12 (24) and II.27.
[95] Made from hogs' lard.

substance every day from the juices of their food. No one should be pleased to share Paracelsus' wild talk; he was clearly fuddled by his own distillations when he attempted to maintain that nutrition occurs only by separation; and that the eye, the nose, the brain and the liver are latent in bread, and roots, leaves and flowers in the moisture of the earth. For just as a craftsman brings out from a crude block of stone or wood leaf, flower, eye, nose, hand, foot and so on by separating off and rejecting what he does not need, so, he claims, *Archaeus*, the Craftsman within bodies, brings out individual limbs and parts by separation and rejection. But joking apart, it is quite certain that the individual parts, both similar and organic, in vegetables and animals, first take in, with some selection, the juices of their foods in almost the same form as each other or with very little difference, and then each assimilates them and turns them into its own nature. Nor does this *assimilation* or *simple generation* occur only in animate bodies, but inanimate bodies also share this process, as has been said of flame and air. Moreover, the non-living spirit, which is contained in every tangible, animate body, is constantly active to digest the heavier parts and turn them into spirit, which would then escape, and that results in loss of weight and drying out, as we have said elsewhere. And in *assimilation* we should also include accretion, which is usually distinguished from nourishment; as when the mud between stones hardens, and turns into a stony material; scale around teeth turns into a substance no less hard than the teeth themselves, etc. For we are of the opinion that there is in all bodies a desire of assimilation, no less than of uniting with homogeneous substances; but this power is constrained, just as the other is, though not in the same ways. We should investigate those ways with all diligence, as well as their dissolution, because they are relevant to the reinvigoration of old age. Finally, it seems to deserve remark that in the other nine motions of which we have spoken,[96] bodies seem only to desire the conservation of their own nature; but in this tenth one their reproduction.

(12) Let the twelfth motion be the motion of *stimulation*; this motion seems to be of the same kind as *assimilation*, and we sometimes give it that name without distinction. It is a motion which is diffusive, communicative, transitive and multiplying, as that is; and they are the same (for the most part) in their effect, though different in manner of producing it and in their subject. For the movement of *assimilation* proceeds as if with power and

---

[96] There are nine motions if the first 'motion' 'common to all matter' is excluded.

authority; it commands and compels the assimilated substance to turn and change into the substance which assimilates it. But the motion of *stimulation* proceeds as by art and insinuation, and with stealth; and merely attracts and adapts the aroused substance to the nature of the substance which arouses it. Also the motion of *assimilation* multiplies and utterly transforms bodies and substances; so there is more flame produced, more air, more spirit, more flesh. But in the motion of *stimulation* only the virtues multiply and pass; and there is more heat produced, more magnetism, more putrefaction. This motion is particularly prominent in heat and cold. For heat does not communicate itself in the process of heating by sharing its primary heat, but only by stimulating the parts of the body to that motion which is the Form of Heat; of which we spoke in the *first harvest of the nature of heat*. And therefore heat is excited in stone or metal much more slowly and with greater difficulty than in air, because these bodies are not adapted and susceptible to that motion. Hence it is likely that there may be inside towards the bowels of the earth materials which altogether refuse to be heated; because on account of their great density they are destitute of the spirit by which that motion is stimulated. Similarly, a magnet imbues iron with a new arrangement of its parts and with a conforming motion, but loses none of its own power. Similarly, leaven, yeast, rennet and some poisons stimulate and excite a successive and continuous motion in a mass of dough, beer, cheese, or in the human body; not so much because of the power of the stimulating body as from the readiness and easy compliance of the body stimulated.

(13) Let the thirteenth motion be the *motion of impression*: this motion too is a kind of *assimilation*, and is the most subtle of the diffusive motions. We have decided to make it a distinct species because of its marked difference from the other two. The simple motion of *assimilation* transforms the actual bodies; so that if you remove the source of the motion, it makes no difference to what follows. For neither the first bursting into flame or the first turning into air has any effect on the flame or air that is subsequently generated. Similarly, the motion of *stimulation* lasts in its full form for a very long time after the source of the motion is taken away; as in a heated body when the source of heat is removed, in aroused iron when the magnet is taken away; and in a mass of dough when the leaven is removed. But though the motion of *impression* is diffusive and transitive, it still seems to depend for ever on the source of its motion; so that if it is taken away or ceases, the motion instantly fails and dies; and therefore it

has its effect in a moment even, or at least in a very short time. And so we have chosen to call the motions of *assimilation* and *stimulation, motions of the generation of Jupiter*, because the generation persists; and the latter the *motion of the generation of Saturn*, because no sooner is it born than it is devoured and swallowed. This motion shows itself in three things: in rays of light, in the striking of sounds and in magnetism so far as its communication is concerned. For as soon as light is taken away, colours and its other images perish; and sound dies very soon after the original stroke and the bodily vibration it causes are finished. For although sounds are buffeted in their flight by the winds as though by waves, still one must carefully note that a sound does not last as long as its reverberation does. For when a bell is struck, the sound seems to continue for a good long time: hence one easily falls into the error of thinking that the sound as it were stays floating in the air all that time; which is absolutely false. For the reverberation is not numerically the same sound but is a repetition. This is made clear by suppressing or stopping the body which was struck. For if the bell stops and is held firmly and doesn't move, the sound immediately stops, and it does not reverberate any more; and if after plucking a string, you touch it again (with a finger in the case of a lute, with a jack in the case of a spinnet), the resonation immediately stops. And if you take away a magnet, the iron instantly falls. But the moon cannot be taken away from the sea, nor the earth from a heavy weight falling. Thus there can be no experiments with them; but it is the same principle.

(14) Let the fourteenth motion be the *motion of configuration* or *position*, by which bodies seem to desire not combination or separation, but *position*, and situation and *configuration* with others. This motion is very obscure, and not well investigated. In some things it seems to be without a cause, though in truth (we think) it is not. For if it is asked why the heaven revolves from east to west rather than from west to east; or why it turns on poles near the Bears rather than on Orion, or some other part of the sky: the question seems to be quite out of order, since those things should be accepted on the basis of experience and as brute facts. And indeed there surely are some things in nature which are ultimate and uncausable; however, this does not seem to be one of them. For we think that the reason for it is a certain harmony and agreement in the universe which has still not yet come under observation. The same questions remain if we accepted that the motion of the earth is from west to east. For it too moves about some poles. And why in the end should those poles be set where they are

rather than anywhere else? Likewise, the polarity, direction and deviation of the compass are attributed to this motion. There are also found in both natural and artificial bodies, especially if they are solid and not fluid, a certain collocation and positioning of parts and what we might call hairs and fibres which need to be carefully investigated, since if we do not find out about them, those bodies cannot be properly discussed or controlled. The motion of *liberty* however is the right place to include those currents in liquids by which, under pressure, they relieve one another in order to distribute the burden equally, until they can free themselves.

(15) Let the fifteenth motion be the *motion of passage*, or the *motion according to the pathways*, by which the powers of bodies are more or less impeded or promoted by the media they are in, in accordance with the nature of the bodies and their active powers, as well as of the medium. For one medium is suited to light, another to sound, another to heat and cold, another to magnetic powers, and others to other powers.

(16) Let the sixteenth motion be the *royal* (as we call it) or *political motion*, by which the dominant and governing parts of a body restrain the other parts, tame, subdue, regulate and compel them to unite, separate, stand, move and take their places in relation to each other, not by their own desires, but with a view to, and in the interests of, the wellbeing of the governing part; so that there is a kind of *government* and *polity* which the ruling part exercises over the subject parts. This motion is most prominent in animal spirit, which tempers all the motions of the other parts so long as it has its strength. It is found also in a lesser degree in other bodies; as has been said of blood and urine which are not dissolved until the spirit which mixed and held their parts together has been expelled or stifled. Nor is this motion simply confined to spirits, though in most bodies spirits dominate because of their swift motion and penetration. However, in dense bodies which are not filled with a strong and lively spirit (such as is found in quicksilver and vitriol), the thicker parts are dominant, so that there is no hope of any new transformation of such bodies unless this curb or yoke is struck off by some art. No one should imagine that we have forgotten the subject under discussion on the ground that (since the sole purpose of this descriptive catalogue of motions is a better investigation, through *instances of struggle*, of their *dominance*) we are now including dominance among the motions themselves. For in the description of the *royal motion*, we are not treating of the *dominance* of motions or powers, but of the *dominance* of parts in bodies. This is the *dominance* which forms this particular species of motion.

(17) Let the seventeenth motion be the *spontaneous motion of rotation*, by which bodies which rejoice in motion and are in a good position take pleasure in their own natures; they pursue only themselves, not other bodies, and try to embrace themselves. Such bodies seem either to move without an end; or to be utterly still; or to move towards an end where, according to their nature, they either rotate or stand still. Bodies in good position which rejoice in motion move in a circle, i.e. in an eternal and infinite motion. Bodies in good position which hate motion, simply rest. Bodies which are not in good position move in a straight line (as the shortest path) to the company of bodies of the same nature. The motion of *rotation* has nine different elements. First, the centre, around which the bodies move; second the poles, upon which they move; third, their circumference or orbit, according to their distance from the centre; fourth, their speed, as they move more swiftly or more slowly; fifth, the direction of their motion, as from east to west or west to east; sixth, their variance from the perfect circle, in spirals which are more or less distant from their centre; seventh, their variance from the perfect circle, in spirals which are more or less distant from their poles; eighth, the shorter or longer distance of their spirals from each other; ninth and last, the variation of the poles themselves, if they are movable; this last has nothing to do with rotation unless it is circular. By a common and ancient belief, this motion is held to be the proper motion of the heavenly bodies. But there is a serious dispute about this motion among some moderns as well as some ancients, who have attributed *rotation* to the earth. But there is another controversy, which is perhaps much more reasonable (if it is not altogether beyond controversy), as to whether (granted that the earth stands still) this motion is confined to the territory of heaven, or whether it comes down and is imparted to the air and the waters. The motion of *rotation* in missiles, however, and in spears, arrows, bullets and so on, we attribute wholly to the motion of *liberty*.

(18) Let the eighteenth motion be the motion of *trembling*. We do not give much credence to this motion, in the astronomers' version of it. But it does come in useful when we make a comprehensive investigation of the appetites of natural bodies; and it seems we should make it a species. It is like a motion of eternal captivity, so to speak. That is, when bodies are not in the position that best suits their nature, and yet are not in a desperate situation, they tremble perpetually, and live in restlessness, neither content with their place, nor daring to go forward. Such a motion is found in the

heart and pulse of animals; and it must exist in all bodies which live thus in an uncertain state between good and bad; under stress they try to free themselves, but then accept defeat, and then renew the attempt again and again.

(19) Let the nineteenth and final motion be the motion to which the name of motion scarcely applies, and yet it really is a motion. We may call it the *motion of rest* or *the motion of horror of motion*. By this motion the earth stands in its own mass, while its extremes move towards the centre; not towards an imaginary centre, but towards union. This is also the appetite by which all highly condensed bodies reject motion; their only appetite is not to move; and though they may be irritated and provoked to move in infinite different ways, they still preserve their nature (so far as they can). And if they are forced to move, they seem to do so simply in order to recover their rest and their position, and not to move again. In this process they show themselves flexible, and they make quite speedy and rapid exertions (as if they were thoroughly fed up, and impatient of any delay). We can only get a partial view of this appetite, because here on earth tangible things are not condensed to the highest degree and are also mixed with some spirit through the warming influence of the heavenly bodies.

And now we have laid out the principles or simple elements of the motions, appetites and active powers which are most widespread in nature. And a good deal of natural science has been outlined in them. We do not claim that no other species could be added; and the divisions themselves might be modified to suit the true lines of things better, and might be reduced to a smaller number. But we do not mean that this is a merely abstract division: as if one were to say that bodies desire their own preservation or growth or reproduction or the enjoyment of their own natures; or that the motions of things tend to preservation and the good, either of the whole, like *indestructibility* and *bonding*; or of the great units, like the motion of *major aggregation, rotation* and *horror of motion*; or of particular forms, like the others. For though these things are true, still unless their matter and structure are defined along true lines, they are speculative, and not much use. However, for the time being they will be adequate, and very useful for weighing *dominances* of powers, and for investigating *instances of struggle*; which is our present subject.

For some of the motions which we have proposed are completely invincible; some are stronger than others and can bind, bridle and control them; some project further than others; some surpass others in time and speed; some nourish others and strengthen, swell and quicken them.

The motion of *indestructibility* is utterly adamantine and invincible. We are still uncertain whether the motion of *bonding* is invincible. And we have not stated for certain that there is a vacuum, whether in empty space or mingled with matter. But we are certain that the reason why the vacuum was introduced by Leucippus and Democritus (namely because without it the same bodies could not enclose and fill spaces of varying size) is false. For matter is like a coil which winds and folds through space, within fixed limits, without the intervention of a vacuum; and there is not two thousand times more vacuum in air than there is in gold (as there would have to be). This is clear enough to us from the powerful virtues of pneumatic bodies (which otherwise would swim in a vacuum like tiny specks of dust), and from many other demonstrations. The other motions rule and are ruled in turn, in proportion to their vigour, quantity, speed and projection, as well as to the assistance or resistance which they meet.

For example: an armed magnet holds and suspends iron sixty times its own weight; so much does the motion of *minor aggregation* prevail over the motion of *major aggregation*; but if the weight is more than that, it gives way. A lever of a certain strength will raise so much weight; to that point the motion of *liberty* prevails over the motion of *major aggregation*; but if the weight is greater, it gives way. Leather stretched to a certain point does not break; to that point the motion of *cohesion* prevails over the motion of *tension*; but if the tension be greater, the leather breaks, and the motion of *cohesion* gives way. Water runs out through a crack of a certain size; to that point the motion of *major aggregation* prevails over the motion of *cohesion*; but if the crack is too small, it gives way, and the motion of *cohesion* prevails. If you put simple sulphur powder into a gun with a ball, and apply fire, the ball is not fired; in this case the motion of *major aggregation* overcomes the motion of *matter*. But if you put gunpowder in, the motion of *matter* in the sulphur prevails, assisted by the motions of *matter* and *avoidance* in the nitre. And so with the rest. For *instances of struggle* (which indicate the *dominance* of powers, and in what amounts and proportions they dominate or give way) are to be sought with keen and constant diligence everywhere.

We also need to make a vigorous investigation of the ways and reasons why motions give way. Do they for example totally cease, or do they struggle to a certain point and then are held in check? For in bodies here on earth there is no true rest, either in wholes or in parts, but only the appearance of it. This apparent rest is caused either by *equilibrium* or by the absolute *dominance* of motions. By *equilibrium* in the case of scales, which stand at rest if the weights are equal. By *dominance* in the case of perforated pots,[97] where the water stays in place and is prevented from falling by the dominance of the motion of *bonding*. However, one should note (as I have said) how far the yielding motions struggle. If a man is pinned down in a fight, stretched out on the ground with arms and legs tied, or otherwise restrained; and yet tries with all his strength to get up, his struggle is no less though he does not succeed. The true situation here (i.e. whether the yielding motion is annihilated by the *dominance* or whether the struggle continues, though not visible) is hidden in conflicts, but will perhaps appear through comparisons. For example: do an experiment with guns to see whether a gun, over the distance that it shoots a ball in a straight line or (as they say) at point-blank range,[98] strikes with less force when shot upward, where the motion of the blow is simple, than it does when shot downwards, where the motion of gravity adds to the force of the blow.

We should also collect the rules of *dominance* which we find: for example that the commoner the good sought, the stronger the motion. Thus the motion of *bonding* involved in the union of the universe is stronger than the motion of gravity involved in the union of dense bodies. Another example is that the appetites which are private goods do not generally prevail against appetites which are more for the public good, except in small quantities. If only this were the case in civil affairs!

XLIX

In the twenty-fifth place among the *privileged instances* we shall put *suggestive*[99] instances, i.e. instances which suggest or point to human benefits. For *being able to* and *knowing* extend human nature in themselves, but do not make it happy. And therefore from the entirety of things we must pick those which do most for human life. But it will be more appropriate to speak of this when we discuss *practical implications*. Moreover in the actual task of *interpreting* individual subjects, we always make room for the *human chart*, or the *wish list*. For both to inquire and to wish appropriately are a part of science.

---

[97] I.e. watering pots with holes in them.
[98] *in puncto blanco*, i.e. the distance the shot travels horizontally before dropping.
[99] *instantiae innuentes*

L

In the twenty-sixth place among *privileged instances* we put *multipurpose instances.*[100] They are instances which are relevant to various topics and come up quite often, and thus save a lot of work and new proofs. There will be a better place to speak of instruments themselves and devices when we discuss applications to practice and methods of experimenting; and those which are already known and in use will be described in the particular histories of individual arts. At present we shall make some general remarks about them simply as examples of *multipurposiveness.*

Man works on natural bodies (apart from simple application and removal) in seven particular ways: by exclusion of obstructive or disturbing objects; by compressing, stretching, shaking and so on; by heat and cold; by keeping a thing in a suitable place; by checking and controlling motion; by special agreements; or by a timely and appropriate alternation, and by a series and succession, of some or all of the above.

(1) For the first: much disturbance is caused by the common air which is all around us and exerting pressure, and by the rays of the heavenly bodies. Devices that seek to exclude them may properly be called *multipurpose*. This is the function of the material and thickness of the vessels in which we place bodies ready to be worked on; also perfect methods of closing vessels, by making them solid and stopping them up with what the chemists call 'putty of wisdom'. A very useful thing is a sealant made by pouring a liquid over a surface, as when they pour oil on wine or herb juices; it spreads over the top like a cover and keeps it well protected from the air. Powders are also quite good; though they have some air in them, they still keep out the force of the open, surrounding air, as when grapes and fruit are preserved in sand and flour. Wax, too, honey, pitch and such viscous substances are properly used as covers and make quite a good seal and keep out air and celestial influences. We have also sometimes tried the experiment of placing a vessel and other bodies within quicksilver, which is by far the densest of all the substances that can be poured around things. Caves and underground cavities are very useful to prevent exposure to the sun and the ravages of the open air; the inhabitants of North Germany use them as granaries. This is also the purpose of keeping things under water; I remember I heard something about someone letting wineskins down into

[100] *instantiae polychrestae*

a deep well (to keep them cool), but then forgetting them and leaving them there, and they stayed there for many years, and when he took them out, the wine had not gone flat or lifeless but was much finer on the palate because (apparently) of a more thorough mixture of its own parts. If the situation requires bodies to be submerged in a depth of water, a river perhaps or the sea, but not to have contact with the water, and not to be shut up in sealed vessels but to be just surrounded by air, very useful is the vessel which is sometimes used to work under the water on sunken ships, which enables divers to stay under water longer and to take breaths in turn from time to time. It was like this. A concave metal barrel was constructed, and was let down evenly into the water, its mouth parallel to the surface; in this way it carried all the air it contained with it to the bottom of the sea. It stood on three feet (like a tripod) which were a little shorter than a man, so that when a diver ran out of breath, he could put his head into the hollow of the jar, take a breath, and then continue with his work. We have heard that a device has just been invented like a small ship or boat, which can carry men under water for a certain distance. Under the kind of jar we mentioned above, certain bodies could easily be suspended; that is why we adduced this experiment.

There is another use of a careful, flawless closure of bodies: namely not merely to prevent the entrance of air from outside (which we have just been discussing), but also to prevent the escape of the spirit of a body which is the subject of an operation inside. For anyone working with natural bodies needs to be certain of his amounts, i.e. be certain that nothing has evaporated or leaked out. For profound alterations occur in bodies when nature prevents annihilation and art also prevents loss or evaporation of any part. On this matter a false belief has become current (and if it were true, there would be virtually no hope of the conservation of a specific amount without loss), i.e. that spirits of substances and air which has been thinned by a high degree of heat cannot be held by any sealant, but seep through the tiny pores in the vessels. Men have been brought to think this by the common experiment of the glass inverted over water with a candle or burning paper inside, as a result of which the water is drawn upwards; and similarly by the experiment of cupping-glasses, which draw up the flesh when warmed over a flame. They think that in both experiments the thinned air is expelled and hence the *quantity* of it is decreased, and therefore the water or the flesh follows by *bonding*. But this is very wrong. For the air is not diminished in *quantity*, but contracted in space; and the consequent movement of the water or flesh only begins with the extinction of the flame or the cooling of the air; so that doctors place sponges soaked in cold water[101] on the cupping-glasses. And therefore there is no reason to be much afraid of an easy escape of air or spirits. For though it is true that even the solidest bodies have their pores, still air or spirit scarcely allows itself to be thinned so fine, just as water refuses to escape through a tiny chink.

(2) Of the second of the seven methods listed, note especially that compressions and such violent forces certainly have the strength to cause motion in space in the most powerful fashion, as in machines and missiles; even to the point of destroying organic body and the virtues which consist wholly in motion. For compressions destroy every kind of life and even every flame and fire, and damage and disable every kind of machine. They also have the power to destroy virtues which consist in arrangement and in a crude difference between elements, as in colours (a bruised flower does not have the same colour as an intact flower, nor crushed amber as whole amber), and in tastes (an unripe pear does not have the same taste as a pear which has been squeezed and worked in the hand, which makes it noticeably sweeter). But these violent forces do not have much effect on the more notable transformations and alterations of similar bodies; because they do not cause bodies to acquire a new solidity that is stable and quiescent but a temporary solidity that is always tending to get free and return to its original form. However, it would not be unprofitable to make some careful experiments in this line, to see whether the condensation or rarefaction of a similar body (such as air, water, oil and so on) when similarly effected by force, could be made stable and fixed, and almost changed in nature. This should first be verified simply by giving it time, and then by the use of instruments and agreements. This would have been easy to do (if I had only thought of it) when I compressed water by hammer and press (as I reported elsewhere),[102] before it burst out. I should have left the flattened sphere for a few days before I let out the water, to find out by experiment whether it would immediately fill the same volume as it had before condensation. If it did not do so, either immediately or shortly after, the condensation could have been clearly seen to be stable; if it did, it would have been apparent that restoration had occurred, and the compression had been temporary. A similar thing should have been done with the air in the glass eggs.[103] I

---

[101] Reading *frigida* for *frigidas*, following Fowler.
[102] II.45.
[103] Cf. II.45.

should have put a firm seal on them immediately after the strong suction; then the eggs should have stayed sealed for some days; and only after that should I have tried to see whether air was drawn through the open hole with a hiss, or whether the same quantity of water would have been drawn in on immersion as there would have been in the beginning if there had been no waiting period. This is probable, or at least worth testing, whether it could and can happen, given that a period of time has similar effects in bodies which are rather more dissimilar. A stick bent by compression fails after a certain time to spring back; this should not be attributed to any loss in the quantity of wood in that time, for the same will happen to a strip of steel (after a longer period), which does not evaporate. But if the experiment does not succeed through simple passage of time, do not abandon the project, but try using some aids. For it is of considerable use if fixed, stable natures can be imposed on bodies by violent forces. In this way air could be changed into water by condensation, and many other such things could be done. For man is more the master of forceful movements than of the others.

(3) The third of the seven methods relates to that great instrument of the operations of nature and of art, that is heat and cold. In this subject human power is plainly lame in one foot. We have the heat of fire, which is infinitely more powerful and more intense than the heat of the sun (as it reaches us) and the heat of animals. But we lack cold, except what can be had in winter or in caves or by packing things in snow and ice, which by comparison can perhaps be equated with the heat of the midday sun in some southern region in the torrid zone when it is intensified by beating off mountains and walls; such heat and cold can be borne by animals at least for a short time. But they are as nothing compared with the heat of a blazing furnace, or the corresponding degree of cold. Therefore all things here among us tend to rarefaction, drying and exhaustion, almost nothing to condensation and thickening except by means of mixtures and artificial methods. And so we must use all diligence to gather instances of cold: such as seem to occur in the exposure of bodies on towers in the bitter cold; in caverns beneath the earth; in packing with ice and snow in deep places dug out for this purpose; in lowering bodies into wells; in covering them in quicksilver and metals; in immersing them in liquids which turn wood into stone; in burying them in the earth (which is said to be the way the Chinese make porcelain, where masses of material suitable for this purpose are said to remain under the earth for forty or fifty years, and to be bequeathed to

heirs, like a kind of artificial mineral); and so on. We should also investigate the condensations which occur in nature which are caused by cold, so that when we have learned their causes, we may apply them to arts: such as we see in the sweating of marble and stones; in the condensation on glass on the inside of windows towards dawn after the cold of the night; in the origin and gathering of mists into subterranean waters, which bubble up as springs; and anything else of this kind.

There are other things that have a cold effect besides things that are cold to the touch; these too have a condensing effect, but seem to work only on the bodies of animals, and hardly on anything else. Many medicines and plasters turn out to be like this. Some condense flesh and tangible parts, for example astringent medicines and congealing medicines; others condense spirits, best seen in sleeping pills. There are two ways of condensing spirits by sleeping pills, or sleep-inducing drugs: one acts by sedation of movement, the other by expulsion of spirits. Violet, dry rose, lettuce and gentle and kindly substances work by means of their friendly and gently cooling vapours to invite the spirits to unite and compose their fierce and anxious motion. Likewise, rosewater placed under the nostrils in cases of fainting refreshes spirits which are too slack and languid, and gives them nourishment. Opiates however and related substances totally expel the spirits by their malign, hostile quality. Hence if they are applied to an external part, the spirits immediately leave that part, and do not easily flow back to it, and if they are taken internally, their vapours rise to the head and completely scatter the spirits contained in the ventricles of the brain; and as the spirits retreat and find no place to escape to, they unite and condense, and are sometimes totally extinguished and smothered. However, these same opiates in moderate doses, by a secondary effect (i.e. the condensation that follows the union), strengthen the spirits and make them more vigorous, and repress their useless, inflammatory motions, as a result of which they contribute a good deal to the cure of diseases and the prolongation of life.

We should also deal with the preparation of substances to receive cold: for example, slightly warm water will freeze more easily than water which is altogether cold, and so on.

Besides, as nature supplies cold so infrequently, we must do as the apothecaries do. When some simple cannot be had, they take a substitute for it, a *quid pro quo*, as they call it: as wood of aloes for balsam, and cassia for cinnamon. In a similar way, we must carefully inquire whether there are any substitutes for cold; i.e. in what ways condensation can be induced in substances other than by cold, which causes them as its own proper effect. There seem to be only four such condensations (so far as is yet clear). The first seems to occur through simple compression, which has little effect on the constant density (for bodies bounce back), but can still be helpful. The second occurs through contraction of the denser parts in a body after the evaporation or escape of the finer parts, as happens when things are hardened by fire or when metals are repeatedly quenched, and so on. The third happens by the coition of homogeneous parts, the most solid parts in any body, which had previously been separated and mingled with less solid parts: as in the restoration of sublimated mercury, which occupies much more space in the form of powder than simple mercury, and likewise in all cleansing of metals from dross. The fourth occurs through agreements, by application of things which cause condensation by a hidden force of the body. Such agreements are as yet barely discernible, which is not surprising since not much is to be expected from an inquiry into agreements until the discovery of forms and structures makes progress. As far as the bodies of animals are concerned, there is no doubt that there are several medicines, taken internally as well as externally, which cause condensation as if by agreement, as we said above. But in inanimate things this kind of effect is rare. Admittedly, there has been quite a noise, both in writing and in rumour, about the story of the tree in one of the islands of the Azores or Canaries (I do not remember which) which drips perpetually, and so gives some supply of water to the inhabitants. And Paracelsus says that a herb called Sun-Dew is filled with dew at midday when the sun is hot and the other grasses around it are dry. We think that both stories are fables, though they would obviously be most remarkably useful, and very well worth investigation, if they were true. Similarly for the sweet dews, like manna, which are found on oakleaves in May: we do not believe that they are caused and condensed by an agreement or property of the oakleaf. Since they also fall on other leaves, they are evidently held and preserved on oakleaves because oakleaves are close knit and not porous, as most other leaves are.

As for heat, man clearly has a wonderful supply abundantly available, and great power over it, but observation and investigation are lacking in some quite vital matters, however the alchemists may boast. For operations involving very intense heat are sought and observed; but those giving gentler heat, which come closest to the ways of nature, are not tried, and therefore escape notice. Hence we see in those furnaces which are most valued that the spirits of bodies are highly excited, as in strong waters and some other chemical oils; tangible parts are hardened, and sometimes fixed, when the volatile element escapes; homogeneous parts are separated; heterogeneous bodies too are incorporated wholesale and merged; and above all the bonds of compound bodies and the subtle structures are confounded and destroyed. They should have tried operations involving a gentler fire and investigated them. More subtle mixtures and orderly structures could be created and derived, on the model of nature and in imitation of the effects of the sun, of the sort of which we sketched some examples in the Aphorism on instances of *alliance*.[104] For the operations of nature are performed with much smaller portions and more precise and discriminating arrangements than in the operations of fire as now applied. Man would truly be seen to increase his authority if by heat and artificial forces operations of nature could be copied in kind, perfected in power and varied in number; to which should be added that they could be speeded up. Rust takes a long time to work on iron, but the effect of sesquioxide appears instantly; similarly with verdigris and white lead. Crystal takes a long time to grow to perfection, but glass is blown in a moment. Rocks take years to form, but bricks are quickly baked, and so on. Therefore (to return to our point) all the different varieties of heat with their respective effects should be diligently and industriously gathered from every source and investigated: celestial heat through rays, direct, reflected, refracted and concentrated in burning-glasses; the heat of lightning, flame, coalfire; fire of different materials; open fire, closed fire, forced fire, raging fire; fire modified by different furnace materials; fire excited by blowing, fire simmering and unstirred; fire at different distances; fire making its way through various media; damp heats, like Mary's baths, dung, the external heat of animals, the internal heat of animals and hay stored in a close place; dry heats, like ashes, lime, warm sand; in fact, every kind of heat with their degrees.

Above all we must attempt to investigate and uncover the effects and operations of the approach and withdrawal of heat by degrees, gradually, regularly, periodically and at specific distances and periods of time. This orderly inequality is truly the daughter of heaven and the mother of generation; no great effect can be expected from violent, sudden or inconsistent heat. This is very obvious even in the case of vegetables; in the wombs of animals too there is great inequality of heat, from the movement, sleep, eating and passions of the females carrying the foetus; finally, in the

---

[104] II.35.

earth's own wombs, the wombs in which metals and fossils are formed, this inequality has its place and power. All the more reason to remark the ineptness of some alchemists of the reformed school, who have thought to achieve their ambitions by means of the constant heat of lamps and such things burning at a steady rate. So much for the operations and effects of heat. This is not the place for a deeper scrutiny before the forms of things and the structures of bodies are further investigated, and come into the light. Once we have firm knowledge of the exemplars, it will be time to seek, devise and adapt instruments.

(4) The fourth mode of operation is through the passage of time, which is the storekeeper and steward of nature, and in a manner the treasurer. We call it passage of time when a body is left to itself for a considerable period, guarded and protected throughout from all external force. For internal motions reveal and perfect themselves when external and adventitious motions cease. The works of age are much more subtle than those of fire. There could not be such a clarification of wine by fire as there is through passage of time; and even incinerations made by fire are less thorough than the dissolution and destruction of the centuries. Instant incorporation and mixing precipitated by fire is much inferior to that effected by the passage of time. The variety of different structures which bodies try through the passage of time (e.g. varieties of putrefaction) are destroyed by fire or a moderately strong heat. It would not therefore be irrelevant to remark that the motions of bodies which have been closely confined have something of violence in them. For the imprisonment impedes the body's spontaneous movements. Therefore the passage of time in an open vessel promotes separation; in a fully closed vessel mixture, in a more or less closed vessel which allows a little air, putrefaction. In any case, instances of the works and effects of the passage of time should be diligently sought from every quarter.

(5) Direction of motion (which is the fifth of the modes of operation) also has great effect. We call it direction of motion when an intervening body impedes, repels, permits and directs the spontaneous movement of another body. It usually consists in the shapes and position of vessels. An upright cone assists the condensation of vapours in alembics; but an inverted cone helps the refining of sugar in receivers.[105] Sometimes bending is required, and narrowing and widening by turns, and so on. It is also the principle in

---

[105] *in vasis resupinatis*, 'receivers' (Ellis)

percolation: an intervening body lets through one element in a substance and holds back another element of it. Percolation or other direction of motion is not always done from outside, but can also be done by means of a body within a body: as when pebbles are put into water to collect the slime; syrups are clarified with the whites of eggs, so that the thicker parts stick to it and may afterwards be separated. Telesio even attributed the shapes of animals to this direction of motion; they were due, he claimed, to the channels and folds of the womb, a gauche and superficial observation. He should have been able to notice a similar formation inside the shells of eggs, where there are no wrinkles or inequality. But it is true that a regulation of motion achieves formation in the case of models and moulds for casting.[106]

(6) Operations by agreements and aversions (which is the sixth mode) are often hidden in the depth. For those so-called occult and specific properties and *sympathies* and *antipathies* are to a great extent corruptions of philosophy. We cannot expect much from discovering the agreements of things before the discovery of the forms and simple structures. For agreement is nothing other than a reciprocal symmetry of forms and structures.

However, the larger and more universal agreements of things are not wholly obscure. And so we must begin from them. Their first and highest distinction is this: certain bodies are quite different from each other in the abundance and rarity of their matter, but agree in structure; others on the other hand agree in abundance and rarity of material, but differ in structure. For it has been well observed by the chemists, in their triad of principles, that sulphur and mercury pervade virtually everything. (Their reasoning about salt is inept and introduced in order to include earthly, dry and fixed bodies.) But in those two at any rate there seems to be visible a kind of natural agreement of the most universal kind. The agreements of sulphur are: oil and fatty vapour; flame; and perhaps star substance. In the other case there is agreement between mercury and water and watery vapours; air; and perhaps the pure interstellar ether. And yet these twin quaternions,[107] or great tribes of things (each within its order), differ immensely in amount of matter and density, but agree closely in structure, as is apparent in many things. On the other hand, different metals agree well in the their abundance and density (especially in comparison with vegetables etc.), but

---

[106] *in modulis et proplasticis*
[107] Two sets of four.

differ in many different ways in structure; and similarly different vegetable and animals vary almost infinitely in structure, but are only a few degrees apart in amount of matter or density.

The next most universal agreement is that of the principal bodies and their sustenance, i.e. the base substances and their nourishment. And so one must inquire in what climates, in what terrain and at what depth individual metals are generated; and similarly of gems, be their origin from rocks or among minerals; in what kind of soil the different trees, bushes and plants grow best and prosper; together with the enrichments which are most helpful, whether manure of various kinds or chalk, seasand, ash etc.; and which of these are most suitable and helpful in which kind of soil. Also heavily dependent on agreement are the planting and grafting of trees and plants, and their various methods, i.e. which plants are most successfully engrafted on which, etc. On this subject it would be a pleasing experiment, which we heard has been recently tried, to engraft forest trees (up to now it has usually been done only with garden trees); the result is that the leaves and nuts are much increased, and the trees give more shade. Similarly, what animals eat should be noted respectively for each kind, and with their negatives. For meateaters cannot survive if fed on herbs; and this is also the reason why (even though the human will has more power over its body than that of the other animals) the Order of the Feuillans[108] almost disappeared after trying the experiment (as is reported), as if human nature could not bear it. We should also note the different materials of putrefaction, from which tiny creatures are generated.

The agreements of principal bodies with their subordinates (for the things we have mentioned may be regarded as such) are quite evident. We may add the agreements of the senses with their objects. Since these agreements are very obvious and well noted and sharply scrutinised, they may throw much light on other, hidden agreements.

The interior agreements and aversions of bodies, or friendships and conflicts (for I am quite tired of the words 'sympathies' and 'antipathies' because of superstitions and stupidities), are falsely associated, or tainted with fables, or little known because ignored. For if anyone asserts that there is conflict between the vine and the cabbage because when planted next to each other they do not do very well, the reason is obvious: both plants are

---

[108] *Folitani*: apparently the Feuillans, French Cistercian monks at the Abbey of Feuillans, who in 1573 began to adopt an exceedingly rigorous rule of life, which led to a number of deaths among them before it was moderated.

aggressive suckers, and rob each other. If anyone asserts that there is agreement and friendship between corn and the cornflower or the wild poppy because these plants grow almost exclusively in cultivated fields, he ought rather to say that there is a conflict between them, because both poppy and cornflower spring and grow from some juice in the soil which the corn has left and rejected; hence the sowing of corn prepares the earth for their growth. There is a great number of such false associations. As for fables, they should be completely exterminated. There remains a very slender store of agreements which have been proved by certain experiment, such as magnet and iron, gold and quicksilver, and so on. Some other noteworthy cases are found in chemical experiments to do with minerals. The commonest of them (such a small number anyway) are found in some medicines, which from their occult and specific properties (as they call them) have a relationship with limbs or humours or diseases or sometimes with individual natures. And we should not omit the agreements between the motions and phases of the moon and the conditions of lower bodies, according as they can be gathered and accepted from experiments in agriculture, navigation and medicine, or from elsewhere with a strict, honest selection. But universal instances of more hidden agreements, the more infrequent they are, the more care is needed for the inquiry by means of reports and faithful and honest narratives, provided this is done without folly or credulity, but with scrupulous and almost sceptical faithfulness. There remains the agreement of bodies which is not artificial in its mode of operation but multipurpose in its application, which we should certainly not neglect but investigate with careful observation. This is the coition or union of bodies, which may be easy or difficult, by compounding or by simple juxtaposition. For some bodies easily and freely mix and incorporate with each other, but others with difficulty and reluctance; for example, powders are best incorporated with waters; limes and ashes with oils; and so on. And we should not only collect instances of bodies' tendency or aversion to being mixed, but also instances of the arrangement of their parts, and of distribution and digestion after they are mixed, and finally also of dominance once the mixing has been completed.

(7) There remains in the last place the seventh and last of the seven modes of operation, i.e. operation in which the other six alternate and take turns. But before we make a deeper inquiry into each one, it would not be appropriate to give examples. It is a thing difficult in thought yet potent in practical effects, to elaborate a series or chain of this kind of alternation as

it applies to individual effects. A supreme impatience possesses and holds men in this kind of thing, both in inquiry and practice; yet it is the thread of the labyrinth as far as major results are concerned. But let this suffice as an example of multipurposiveness.

LI

In the twenty-seventh and last place among *privileged instances* we shall put *magical instances.*[109] By this name we mean instances in which the matter or the efficient cause is slight or small in comparison with the effect or result which follows. So that even if they are common, they are still like a miracle, some at first glance, others even after attentive observation. Nature supplies these herself sparingly; it will appear in future times what she will do when her lap is shaken out, after the discovery of forms, processes and structures. But (so far as we conjecture in our time) these *magical effects* happen in three ways. They happen first by self-multiplication, as in fire, and the so-called specific poisons, and also in movement communicated and strengthened from one wheel to another. Or they happen by excitement or attraction in another object, as in a magnet which excites a large number of needles without losing or lessening any of its own virtues, and in yeast and such things. They happen thirdly by anticipation of a motion, as noted in the case of gunpowder, cannon and mines. The first two means require an investigation of agreements, the third the measuring of motions. We have as yet no sound indications whether there is any way of changing bodies through their smallest parts, or 'minima' (as they call them),[110] and of transforming the subtler structures of matter (which occurs in every kind of transformation of matter, so that art may do in a short time what nature achieves through many windings). And as we aim in what is solid and true to achieve our final and highest goals, so we consistently despise the vain and presumptuous, and do our best to get rid of it.

LII

So much for *privileged instances,* or *first-class instances.* I should add the reminder that in this *Organon* of ours we are dealing with logic, not philosophy. But our logic instructs the understanding and trains it, not (as

[109] *instantiae magicae*
[110] *per minima (ut vocant)*

common logic does) to grope and clutch at abstracts with feeble mental tendrils, but to dissect nature truly, and to discover the powers and actions of bodies and their laws limned in matter. Hence this science takes its origin not only from the nature of the mind but from the nature of things; and therefore it is no wonder if it is strewn and illustrated throughout with observations and experiments of nature as samples of our art. The *privileged instances* (as is clear from our account) are twenty-seven in number; and are: *solitary instances; instances of transition; revealing instances; concealed instances; constitutive instances; instances of resemblance; unique instances; deviant instances; borderline instances; instances of power; instances of association and aversion; accessory instances; instances of alliance; crucial instances; instances of divergence; instances that open doors or gates; summoning instances; instances of the road; instances of supplement; cleaving instances; instances of the rod; running instances; doses of nature; instances of struggle; suggestive instances; multipurpose instances; magical instances.* The use of these *instances*, in which they surpass ordinary instances, tends in general either in the direction of information or in the direction of operation, or in both. In the informative aspect they assist either the senses or the understanding. In the case of the five *instances of the lamp*, for example, they assist the senses.[111] They assist the understanding either by speeding the exclusion of a form, as *solitary instances* do; or by narrowing and closely delimiting the affirmation of a form, as do *instances of transition, revealing instances* and *instances of association*, as well as *accessory instances*; or by raising the understanding and guiding it to general and common natures: which they do either directly, as do *concealed and unique instances* and *instances of alliance*; or in a high degree, as do *constitutive instances*; or only slightly, as do *instances of resemblance*; or by correcting the understanding from its habitual channels, as do *deviant instances*; or by guiding it to the great form or structure of the whole, as do *borderline instances*; or by warning of false forms and causes, as do *crucial instances* and *instances of divergence*. As for the practical aspect, *privileged instances* either designate, measure or facilitate practice. They designate either by pointing to where we should begin so as not to repeat what has already been done, as do *instances of power*; or what we should aim at if we have the opportunity, as do *suggestive instances*; the four *mathematical instances* measure;[112] the facilitating instances are the *multipurpose instances* and the *magical*.

---

111 The five types of instance described in 11.39–43 are introduced in 11.38 as 'instances of the lamp'.
112 The four 'mathematical instances' are nos. 21–4 in 11.45–9.

Again, we should make a collection of some of the twenty-seven instances now at the beginning (as we said above of some of them), and not wait for a special inquiry into natures. Such are the *instance of resemblance, the unique, the deviant and the borderline instances, the instances of power, the instances that open doors or gates, the suggestive, the multipurpose* and *the magical instances.* For these either help or heal the understanding and the senses, or assist practice in general. The rest should be sought when we draw up *tables of presentation* for the purpose of interpreting a particular nature. For *instances* endowed and distinguished by *these privileges* are like the soul among the ordinary instances of presentation; and as we said at the beginning, a few of them are as good as many; hence when we make *tables*, we should investigate them with great vigour, and put them in *tables.* We shall also have to speak of them in what follows, and that is why we had to put their treatment first.

And now we must proceed to the *aids and corrections of induction*, and after that to *concrete things*, and to *latent processes* and *latent structures*, and the other things which we set out in due order in Aphorism 21. We intend at the end (like honest and faithful guardians) to hand men their fortunes when their understanding is freed from tutelage and comes of age, from which an improvement of the human condition must follow, and greater power over nature. For by the Fall man declined from the state of innocence and from his kingdom over the creatures. Both things can be repaired even in this life to some extent, the former by religion and faith, the latter by the arts and sciences. For the Curse did not make the creation an utter and irrevocable outlaw. In virtue of the sentence 'In the sweat of thy face shalt thou eat bread',[113] man, by manifold labours (and not by disputations, certainly, or by useless magical ceremonies), compels the creation, in time and in part, to provide him with bread, that is to serve the purposes of human life.

End of the Second Book of *The New Organon.*

[113] Genesis 3:19.

Outline
of a Natural and Experimental History,
adequate to serve as
the basis and foundation of
True Philosophy

The reason why we are publishing our *Renewal* in parts is that some of it may be put out of danger. For the same reason we are prompted to append a small section of the work at this point and to publish it with what we have completed above. This is the outline and sketch of a Natural and Experimental History such as is appropriate for grounding a philosophy; it contains an abundance of good material, which is digested for the work of the interpreter which follows it. The proper place for this would be when we duly reached the *Preparations* of Investigation. But we think it advisable to anticipate, and not wait for the proper place; because the kind of history we have in mind and are about to describe is a massive thing which would take great pains and expense to complete; it requires the efforts of many men, and as we have said before, is in some sense a royal task. It occurs to me therefore that it would be appropriate to see whether there may be others able to take up this challenge, so that while we complete the whole work according to plan, this complex and time-consuming part may be built and made available (if it shall please the divine majesty) even in our lifetime, by the cooperation of others working steadily along with us, especially as our own resources without help would hardly be adequate for such a great province. We shall perhaps succeed in completing by our own efforts the part which relates to the actual work of the understanding. But the materials for the understanding are so widely scattered that we need to have agents and merchants (so to speak) seeking and collecting them from every corner. And in fact it is rather below the dignity of our enterprise to waste our own time on such a thing as any

---

[1] Published in 1620 in the same volume with *The New Organon*.

industrious person may do. We will now present the main point of the matter ourselves, and give a careful and exact account of the method and outline of a such a history as will be adequate to our design, so that men may be instructed, and not continue to be guided by the example of the currently available natural histories, and stray far from our design. Meanwhile we should stress here what we have often said elsewhere, that if the whole human race had dedicated itself and its efforts to philosophy, and the whole earth had been, or should become, absolutely filled with universities and colleges and schools of learned men, they could not have made, and cannot make, any progress in philosophy and the sciences worthy of the human race without such a Natural and Experimental History as we shall now prescribe. On the other hand, when such a history has been developed and built up well, with the ancillary and illuminating experiments which will occur or will have to be devised in the actual process of interpretation, the investigation of nature and the sciences will be the work of a few years. This is what has to be done, or the enterprise abandoned. It is the one and only method by which a true and practical philosophy can be established; and men will then perceive, as if awakening from a deep sleep, what is the difference between the opinions and fictions of the mind and a true and practical philosophy, and just what it is to consult nature herself about nature.

Therefore we shall first give general instructions for compiling such a history; then we shall set before men's eyes its particular form, with occasional remarks on *the purpose* to which the inquiry has to be fitted and adapted as well as its *subject*, so that when the scope of the thing is properly understood and envisaged, it may bring other things into men's minds which perhaps we have missed. We have chosen to call this history the *Primary History* or *Mother History*.

Aphorisms
On Compiling a Primary History

Aphorism I

Nature exists in three states and accepts three kinds of regime. She is either free and unfolding in her own ordinary course, or driven from her state by the vicious and insolent assaults of matter and by the force of obstructions, or constrained and shaped by human art and agency. The first state refers to the species of things, the second to prodigies, the third to artificial things. For in artificial things nature accepts the yoke from the empire of man; for these things would never have been done without man. A completely new face is given to bodies by human effort and agency, a different universe of things, a different theatre. There are, consequently, three forms of natural history. It deals either with the *Freedom* of nature or with the *Errors* of nature or with the *Bonds* of nature; so that a good division we might make would be a history of *Births*, a history of *Prodigious Births*, and a history of *Arts*; the last of which we have also often called the *Mechanical* and the *Experimental Art*. But we are not prescribing that the three should be treated separately. Why may not the histories of prodigies in particular species be rightly joined with the history of the species themselves? Artificial things too are sometimes rightly joined with species, and sometimes better separated. It will be best therefore to decide this in each individual case. For excess of method and lack of method equally give rise to repetitions and prolixity.

II

Natural history, as we have said, has three subjects, but two uses. It is used either for knowledge of the things which are committed to the history, or it is used as the first matter of philosophy and the stuff and material of true induction. The latter use is now under discussion; now, I say, and never ever before this. For neither Aristotle nor Theophrastus nor Dioscorides nor Caius Plinius,[2] much less the moderns, ever suggested this purpose (of which we speak) for natural philosophy. The main point is that those who take on the role of writing history hereafter should constantly reflect and bear in mind that they are not to serve the pleasure of the reader nor the immediate advantage which can be got from reports, but must find and build a store of things sufficiently large and varied to formulate true axioms. If they keep this in mind, they will prescribe for themselves the means of such a history. For the end governs the means.

---

[2] Aristotle (384–322), Greek philosopher and investigator of nature, founded the Lyceum as a school of philosophy and research; Theophrastus (*c.* 372–287 BC) succeeded him as head of the Lyceum and is particularly known for his *History of Plants*. Dioscorides (a physician of the 1st century AD) wrote a *Herbal* which remained standard for centuries. C. Plinius (= 'Pliny the Elder', AD *c.* 23–79) wrote *Historia naturalis*, an encyclopedia of the natural science of his day, and famously died while attempting to study the eruption of Vesuvius in AD 79.

III

But the more effort and labour this enterprise entails, the less appropriate it is to load it down with irrelevancies. Men need to be clearly warned against putting too much effort into three things which enormously increase the amount of work, but add little or nothing to its quality.

First, then, they must do without antiquities and citations of authors and authorities; also disputes, controversies and dissenting opinions – in a word, philology. Do not cite an author except in a matter of doubtful credit; do not introduce a controversy except in a case of great importance. Reject everything that makes for ornament of speech, and similes, and the whole repertoire of eloquence, and such vanities. State all the things you accept briefly and summarily, so that there may be no more words than there are things. For no one who collects and stores materials for buildings or ships or such structures places them prettily (like window-dressers) and shows them off to please, but only makes sure that they are good and sound, and take up the least space in the warehouse. That is just the way it should be done here.

Secondly, there is little point in natural histories, indulging in numerous descriptions and pictures of species and in minute varieties of the same things. Such petty variations are nothing more than nature's fun and games, and are quite close to the nature of an individual. They offer a kind of ramble through the things themselves which is attractive and delightful, but give little information for the sciences, and what they do give is more or less superfluous.

Thirdly, we have to bid a stern farewell to all superstitious stories (I do not say stories of prodigies when the memory of them is reliable and probable, but superstitious stories), and experiments of ritual magic. We do not want the infancy of philosophy, which gets its first breast from natural history, to become used to old women's stories. There will perhaps be time (after we have penetrated a little more deeply into the investigation of nature) for skimming through them, so that if there is any natural virtue in those dung-heaps, it may be extracted and put to use. Meanwhile they should be kept away. Also experiments of natural magic should be carefully checked and severely critiqued before being accepted, especially those which are commonly derived from vulgar *sympathies* and *antipathies*, a very idle practice that depends on a combination of facile credulity and imaginative invention.

We have already achieved a good deal by ridding natural history of these three superfluous things (just mentioned), which otherwise would have filled volumes. But this is not the end. In a great work it is equally necessary to describe what is accepted succinctly as it is to cut out superfluities, though it is evident that such purity and brevity will give much less pleasure to reader and writer alike. We must constantly repeat the point that we are merely building a warehouse or storage space; not a place in which one is to stay or live with pleasure, but which one enters only when necessary, when something has to be taken out for use in the work of the Interpreter which follows.

IV

In the history which we seek and intend, we must be sure above all that it is extensive and made to the measure of the universe. The world must not be contracted to the narrow limits of the understanding (as it has been heretofore), but the understanding must be liberated and expanded to take in the image of the world as it is found to be. The habit of *looking only at a few things and of giving a judgement on the basis of a few things* has ruined everything. And so we take up the division of natural history which we made just now (of Births, Prodigious Births and Arts), and assign five parts to the history of births. The first will be of the ether and the heavens. The second, of the sky[3] and the regions (as they call them) of the air. The third, of earth and sea. The fourth, of the *elements* (as they call them) of flame or fire, air, water and earth. We want elements to be understood in the sense not of the prime qualities of things, but of the major constituents of natural bodies. For the nature of things is so distributed that the quantity or mass of certain bodies is very great, because their structure requires the texture of an easy and common material; such are the four substances which I mentioned. But the quantity of certain other bodies in the universe is small and occurs rarely, because the texture of their matter is very different, very subtle and for the most part delimited and organic; such are the species of natural things, metals, plants, animals. For this reason we have chosen to call bodies of the first kind *major associations*, and the latter *minor associations*. The major associations come into the fourth part of history, under the name of elements, as we have said. I am not confusing the fourth part with the second or third parts simply because in each of them I have mentioned

---

[3] *Meteora*, a Greek word, refers, in Bacon, both to the regions of the sky below the heavens and to the bodies in the sky (some of which are meteors in our sense).

air, water and earth. For in the second and third parts I have given the histories of them as integral parts of the world and so far as they contribute to the fabric and structure of the universe; but the fourth contains the history of their substance and nature, which flourishes in the similar parts of each of them, but does not relate to the whole. Finally, the fifth part of the history contains the minor associations or species, with which natural history to this point has been particularly concerned.

As far as concerns the history of Prodigious Births, we have already said that it goes best with the history of Births; we are referring to history which is natural as well as prodigious. For we insist on relegating the superstitious history of miracles (of whatever kind) to a special treatise of its own; nor should it be started right at the beginning, but a little later, when we have penetrated more deeply into the investigation of nature.

We establish three kinds of history of Arts and of nature changed and altered by man, or Experimental history. Either it is drawn from the Mechanical Arts; or from the practical part of the liberal sciences; or from several practices and experiments which do not form an art of their own, and which in fact sometimes occur as the result of the lowest kind of experience and do not aspire to form an art.

And so once a history has been compiled from all the sources that I have mentioned, Generation, Prodigies, Arts and Experiments, nothing is omitted which may equip the senses to give information to the understanding. And we shall no longer be leaping around in little circles (like dervishes), but in our progress we shall walk the boundaries of the world.

V

The most useful of the parts of history which we have mentioned is the history of arts; it shows things in motion, and leads more directly to practice. It also lifts the mask and veil from natural things, often hidden and obscured by a variety of shapes. And the manipulations of art are like the bonds and shackles of Proteus, which reveal the ultimate strivings and struggles of matter. For bodies refuse to be destroyed or annihilated, but shift into various other shapes. Therefore we must put aside our arrogance and scorn, and give our full attention to this history, despite the fact that it is a mechanics' art (as it may seem), illiberal and mean.

The preferable arts are those which present, alter and prepare natural bodies and the materials of things, like agriculture, cookery, chemistry, dyeing, the manufacture of glass, enamel, sugar, gunpowder, fireworks, paper and suchlike. Of less value are things which essentially consist in the subtle motion of hands and tools, such as weaving, carpentry and metalwork, building, the manufacture of mill-wheels, clocks and so on; though they are certainly not to be ignored either, both because many things occur in them which relate to the alterations of natural bodies, and because they give accurate information about local motion, which is of the highest importance for many things.

In the whole body of this history of the Arts, one piece of advice above all needs to be given and taken to heart: we must accept not only experiments which are relevant to the purpose of the art, but any experiments which happen to come up. For example, locusts or crabs when cooked turn red (before cooking they have the colour of mud); this has nothing to do with preparing a meal; but it is quite a good instance for the investigation of redness, since the same thing happens in baked bricks. Similarly, meat is more quickly salted in winter than in summer; this not only warns the cook to preserve his foods properly and as much as required, but is also a good instance for indicating the nature and effect of cold. For this reason it is a cardinal error (to use an expression) to think that our plan is being followed if experiments of art are collected with the sole aim of improving the individual arts. For although in many cases we do not altogether condemn this, our plan however actually is that the streams of all mechanical experiments should run from all directions into the sea of philosophy. It is on the basis of the privileged instances that we should select the more significant instances in every kind (and we must seek them out and track them down with every care and effort).

VI

We should also summarise here what we treated at greater length in Aphorisms 99, 119 and 120 of the First Book. Here it may be enough to give a brief command in the form of an instruction: accept into this history, first, the commonest things which one might think it superfluous to put into writing because they are so familiar; then mean things, illiberal, disgusting (for *all things are pure to the pure*, and if the tax receipts from urine[4]

---

[4] The emperor Vespasian placed a tax on urine. When his son Titus protested, Vespasian held a coin derived from this tax before his nose and asked him if it smelt. Titus had to admit that it did not. 'And yet it comes from piss', said Vespasian (Suetonius, *The Twelve Caesars*, 'Life of Vespasian', 23).

smelt good, much more so is the light and information we may get from any source); also, trivial, childish things (and no wonder, for we must become again as children, utterly);[5] finally, things which seem to be excessively subtle, because they are of no use in themselves. For (as I have said) the things we have exhibited in this history have been collected for their own sakes; and therefore it is not fair to measure their dignity in themselves, but so far as they can be transferred to other things, and contribute to philosophy.

VII

We also prescribe that all things in both natural bodies and natural powers be (as far as possible) numbered, weighed, measured and determined. For we are planning works, not speculations. And a proper combination of physics and mathematics yields practical results. For this reason we must investigate and we must describe, in the history of heavenly things, the exact returnings and distances of the planets; in the history of land and sea, the extent of land and how much of the surface it occupies compared with waters; in the history of air, how much compression it permits without powerful resistance; in the history of metals, how much heavier one metal is than another; and countless other things of this kind. When exact measurements are not to be had, then we must certainly have recourse to unspecified estimates and comparisons: for example, if perhaps we lack confidence in the astronomers' calculations of distances, that the moon is within the shadow of the earth, that Mercury is above the moon, and so on. And when we cannot make intermediate measurements, we should set out the measurements at the limit: for example, that a fairly weak magnet lifts iron to a certain weight, in relation to the weight of the stone itself; and that the most powerful magnet raises as much as sixty times its own weight; as we have seen happen in the case of a very small armed magnet. We know well enough that these determinate instances do not often or easily occur, but need to be sought out as aids (when the situation most demands it) in the actual course of investigation. If however they should occur by chance, they should be included in the natural history if they do not too much delay the process of completing it.

---

[5] Cf. Matthew 19:14.

VIII

As for the reliability of what should be accepted into a history: they are necessarily either absolutely reliable, dubiously reliable or not reliable at all. Things of the first kind we should simply report; things of the second kind we record with a remark, i.e. with the words 'it is reported, or they say, or I have heard from a reliable source', and so on. It would be too tedious to include the arguments on both sides, and will certainly hold the writer up too long. And it makes little difference to the question under discussion, since (as we said in Aphorism 118 of Book I) the false experiments, if not too prevalent, will soon be shown up by true axioms. However, if the instance is very notable, either in the use itself or because many other things may depend on it, then one certainly has to give the name of the author, and not just his name, but with a remark as to whether his statement is based on a report or description (as is nearly always the case in the writings of Caius Plinius) or on his own knowledge; also whether the thing happened in his own time or in the past; and whether it was the sort of thing that requires many witnesses to be true; finally, whether the author has shown himself to be boastful and frivolous or strict and sober; and similar points which go to the question of reliability. Finally, there are things which are completely unreliable and yet widely and popularly believed, things which have stayed current for centuries, partly through neglect and partly through the use of analogies (e.g. that a diamond binds a magnet, garlic takes away its power, amber attracts everything except basil, and so on); it is not appropriate simply to ignore these; we have to ban their use in so many words, so that they will no longer trouble the sciences.

It would also be useful to note the origin of any foolish or credulous belief if one should come across it, the belief, for example, that the herb savory has the power of exciting desire for the simple reason that its root has the shape of testicles. The truth is that it has this shape because a new bulb is made every year while the previous year's bulb is still in place, and that is why it is double. This is evident from the fact that the new root is found to be solid and juicy, while the old one is withered and spongy. So it is not surprising that one sinks in water while the other floats; yet this is thought to be a marvel, and has given credence to the herb's other powers.

IX

There are some useful additions to a natural history which may make it more fit and helpful to the subsequent work of an interpreter. They are five. First, questions (not of causes but of fact) should be added, to encourage and provoke further investigation; for example, in the history of land and sea, whether the Caspian sea has tides, and at what intervals; whether there is a Southern continent, or only islands; and so on.

Secondly, in any new experiment of any subtlety, we should append the actual method used in the experiment, so that men may have the opportunity to judge whether the information it produced is reliable or deceptive, and also to encourage men to apply themselves to look for more accurate methods (if there are any).

Thirdly, if there is anything doubtful or questionable in any account, we are wholly against suppression or silence about it; a full and clear note should be attached as a remark or warning. We want the first history to be composed with utter scrupulousness, as if an oath had been taken about the truth of every detail; for it is the volume of the works of God, and (so far as one may compare the majesty of the divine with the humble things of the earth) like a second Scripture.

Fourth, it would not be out of place to insert occasional observations (as C. Plinius did); for example, in the history of land and sea, that the shape of the lands (so far known) in relation to the seas is narrow and almost pointed to the south, wide and broad to the north; the shape of the seas is the opposite; and the great oceans cut through the lands in channels that run north–south rather than east–west, except perhaps in the furthest polar regions. Also it is a good thing to add canons (which are simply general and universal observations); for example, in the history of heavenly bodies, that Venus is never more than 46 degrees from the sun, Mercury, 23; and that the planets which lie above the sun move very slowly, since they are furthest from earth and the planets below the sun very fast. Another kind of observation to make, which has never yet been used, despite its importance, is this: to append to an account of what is a mention of what is not. For example, in the history of heavenly things, that no star is found to be oblong or triangular, but that every star is globular: either simply globular like the moon, or spiky in appearance but globular at the centre, like the other stars, or apparently shaggy but globular at the centre, like the sun; or that the stars are scattered about in no order at all, so that there is no such thing among them as a quincunx or a quadrangle or any other perfect figure (despite the names they have been given, like delta, crown, cross, chariot etc.); hardly even a straight line, except perhaps in the belt and dagger of Orion.

Fifthly, something which quite depresses and destroys a believer will perhaps help an investigator: namely to survey in a brief and summary form of words the currently accepted opinions in all the variety of the different schools; enough to wake up the intellect and no more.

X

This is enough for general precepts. If they are carefully observed, the historical task will go straight towards its purpose, and will not get too big. But if it seems to some faint-hearted person to be an immense task, even in this circumscribed and limited form, let him turn his eyes to the libraries; and among other things, let him look on one side at the texts of the civil and canon law, and on the other at the commentaries of the doctors and the jurists on them; and let him see the difference between them with regard to bulk and volume. Brevity suits us; like faithful scribes, we pick up and write down the very laws of nature, and nothing else; brevity is almost imposed by things themselves. Opinions, dogmas and speculations, however, are innumerable; there is no end to them.

In the Plan of our Work we mentioned the *cardinal powers* in nature, and that one must compose their history before approaching the task of interpretation. We have not forgotten this, but have kept it for ourselves to do, since we dare not rely fully on other men's labour in this area until they have begun to develop a rather closer familiarity with nature. Now therefore we should move on to an outline account of *Particular Histories*.

But as we are now distracted by business, we have only the time to append a Catalogue of titles of Particular Histories. As soon as we have the leisure for the task, we plan to give detailed instructions by putting the questions that most need to be investigated and written up in each history because they help to fulfil our purpose, like certain particular *Topics*. Or rather (to use the language of civil procedure) we intend, in this *Great Suit* or *Trial*, given and granted by the goodness and providence of God (by which the human race seeks to recover its right over nature), to cross-examine nature herself and the arts on the articles of the case.

CATALOGUE OF
PARTICULAR HISTORIES
BY TITLES

1. History of the Heavens; or Astronomy.
2. History of the Structure of the Sky and of its parts towards the Earth and its parts; or Cosmography.
3. History of Comets.
4. History of Blazing Meteors.
5. History of Lightning, Thunderbolts, Thunder and Sheet-lightning.
6. History of Winds, and Sudden Blasts, and Waves of Air.
7. History of Rainbows.
8. History of Clouds, as seen above.
9. History of Blue Sky, Twilight, multiple Suns, multiple Moons, Haloes, various Colours of the Sun and Moon; and of every variation in the appearance of the heavenly Bodies caused by the medium.
10. History of Normal Rain, Storms and Abnormal Rains.
11. History of Hail, Snow, Ice, Frost, Fog, Dew and so on.
12. History of all other things that fall or come down from above, and are generated above.
13. History of Sounds above (if there are any) except Thunder.
14. History of Air in general, or in the Structure of the World.
15. History of the Weather or Temperatures of the Year, both by the differences of Regions, and by the characteristics of the Times and periods of the Year; of Floods, Hot Spells, Droughts and so on.
16. History of Land and Sea; of their Shape and Extent and their Structure in relation to each other, and of their Extent in breadth or narrowness; of Land Islands in the Sea, of Gulfs of the Sea, and of salt Lakes on Land, of Isthmuses, Promontories.
17. History of Motions (if there are any) of the globe of Land and Sea; and of the Experiments by which they may be inferred.
18. History of the major Motions and Disturbances in Land and Sea; namely Earthquakes, Tremors and Fissures, emergence of newly formed Islands, floating Islands, Loss of Land by encroachment of the Sea, Inundations and Floods, and by contrast Withdrawal of the Sea;

Eruptions of Fire from the Earth, sudden Eruptions of Water from the Earth, and the like.

19. Natural Geographical History, of Mountains, Valleys, Forests, Plains, Deserts, Marshes, Lakes, Rivers, Torrents, Springs, and all their different ways of arising, and the like; omitting Nations, Provinces, Cities and such Civil matters.

20. History of the Tides of the Sea, Currents, Swells, and other Motions of the Sea.

21. History of the other Qualities of the Sea: its Salinity, various Colours, Depth; and of Rocks, Mountains and Valleys under the sea, and suchlike.

There follow the Histories of the Major Masses

22. History of Fire and of Burning Things.

23. History of Air, as a Substance, not in Configuration.

24. History of Water, as a Substance, not in Configuration.

25. History of Earth, and its diversity, as a Substance, not in Configuration.

There follow the Histories of Species

26. History of the perfect Metals, Gold, Silver; and their Ores, Veins and Pyrites; also Workings in their Ores.

27. History of Quicksilver.

28. History of Fossils;[6] as Vitriol, Sulphur etc.

29. History of Gems; as Diamond, Ruby etc.

30. History of Stones; as Marble, Quartz,[7] Flint etc.

31. History of the Magnet.

32. History of Miscellaneous Bodies which are neither quite Fossil nor Vegetable; as Salts, Amber, Ambergris etc.

33. Chemical History of Metals and Minerals.

34. History of Plants, Trees, Shrubs, Herbs: and of their Parts, Roots, Stems, Wood, Leaves, Flowers, Fruits, Seeds, Gums etc.

35. Chemical History of Vegetables.

36. History of Fish, and of their Parts and Generation.

37. History of Birds, and of their Parts and Generation.

38. History of Quadrupeds, and of their Parts and Generation.

[6] In the older sense of 'any rock, mineral or mineral substance dug from the earth' (OED).
[7] *Lapis Lydius*

39. History of Snakes, Worms, Flies and other Insects; and of their Parts and Generation.
40. Chemical History of Animal Products.

There follow the Histories of Man

41. History of the Figure and external Members of Man, his Size, Frame, Face and Features; and of their variations by People and Climate, or other minor differences.
42. Physiognomic History of these.
43. Anatomical History, or History of the internal Members of man; and of their variety as it is found in the natural Frame and Structure, and not merely with regard to Diseases and Abnormal Features.
44. History of the common parts of Man; as of Flesh, Bones, Membranes etc.
45. History of the Fluids in Man: Blood, Bile, Sperm etc.
46. History of the Excrements: Saliva, Urine, Sweat, Stools, Head-hair, Body-hair, Hang-nails, Nails and so on.
47. History of the Functions: Attraction, Digestion, Retention, Expulsion, Blood formation, Assimilation of food into the members, Conversion of Blood and its Flower into Spirit, etc.
48. History of the Natural and Involuntary Motions; as of the Motion of the Heart, the Motion of the Pulses, Snoring, Lungs, Erection of the Penis etc.
49. History of the mixed Motions, which are both natural and voluntary; as of Breathing, Coughing, Urination, Excretion etc.
50. History of the Voluntary Motions; as of the Instruments of articulate speech; Motions of the Eyes, Tongue, Jaws, Hands, Fingers; of Swallowing etc.
51. History of Sleep and Dreams.
52. History of different Conditions of the Body; Fatness, Thinness; of the so-called Complexions,[8] etc.
53. History of Human Birth.
54. History of Conception, Quickening, Gestation in the Womb, Birth etc.
55. History of Human Nourishment, and of everything that is eaten and drunk, and of every Diet; and of their Variation between peoples or minor differences.

---

[8] Combinations of the four humours: hot and cold, wet and dry.

56. History of the Growth and Increase of the Body as a whole and in its parts.
57. History of Course of Life: of Infancy, Childhood, Youth, Old Age, Longevity, Brevity of Life and so on, by peoples, and minor differences.
58. History of Life and Death.
59. Medical History of Diseases and their Symptoms and Signs.
60. Medical History of the Treatment, Remedies and Cures of Diseases.
61. Medical History of things which preserve Body and Health.
62. Medical History of things pertaining to the Form and Beauty of the Body.
63. Medical History of things which change the Body, and pertain to Control of Change.
64. Pharmaceutical History.
65. Surgical History.
66. Chemical History of Medicines.
67. History of Sight and Visible things, or Optics.
68. History of Painting, Sculpture, the Plastic Arts etc.
69. History of Hearing and Sounds.
70. History of Music.
71. History of Smell, and Odours.
72. History of Taste, and Tastes.
73. History of Touch, and its Objects.
74. History of Sex, as a species of Touch.
75. History of bodily Pains, as a species of Touch.
76. History of Pleasure and Pain in general.
77. History of the Passions; as of Anger, Love, Shame etc.
78. History of the Intellectual Faculties: Thought, Imagination, Discourse, Memory etc.
79. History of Natural Divinations.
80. History of Diagnosis, or of concealed Natural Distinctions.
81. History of Cooking, and of related arts, such as those of the Butcher, Poulterer etc.
82. History of Baking and Breadmaking, and related arts, like Milling etc.
83. History of Wine.
84. History of the Cellar, and of the different kinds of Drink.
85. History of Cakes and Confections.
86. History of Honey.
87. History of Sugar.

88. History of Dairy products.
89. History of the Bath, and of Ointments.
90. Miscellaneous History of the Care of the Body; Hairdressing, Perfumes etc.
91. History of Goldworking, and related arts.
92. History of Woolworking and related arts.
93. History of Manufactures from Satin and Silk, and related arts.
94. History of Manufactures from Linen, Hemp, Cotton, Hair and other fibres, and related arts.
95. History of Feather goods.
96. History of Weaving and related arts.
97. History of Dyeing.
98. History of Tanning, Leatherworking and related arts.
99. History of Bed-stuffing and Cushions.
100. History of Ironworking.
101. History of Quarrying or Stonecutting.
102. History of Brick- and Tile-making.
103. History of Pottery.
104. History of Cementing and Plastering.
105. History of Woodworking.
106. History of Leadworking.
107. History of Glass and of all glassy Substances and of Glassmaking.
108. History of Architecture in general.
109. History of Wagons, Chariots, Litters etc.
110. History of Printing, Books, Writing, Sealing; of Ink and Pen, of Paper, Parchment etc.
111. History of Wax.
112. History of Wickerwork.
113. History of Matmaking, and of Manufactures of Straw, Rushes and so on.
114. History of Washing, Sweeping etc.
115. History of Agriculture, Pasturing, Forestry etc.
116. History of Gardening.
117. History of Fishing.
118. History of Hunting and Hawking.
119. History of the Art of War, and related arts: the arts concerned with Armoury, Bows, Arrows, Guns, Artillery, Catapults, Siege-engines etc.

120. History of the Naval Art, and the related techniques and arts.
121. History of Athletics and of Human Exercise of every kind.
122. History of Horsemanship.
123. History of all kinds of Games.
124. History of Jugglers and Clowns.
125. Miscellaneous History of different Artificial Materials; such as Enamel, Porcelain, the various Cements etc.
126. History of Salts.
127. Miscellaneous History of different Machines, and Motions.
128. Miscellaneous History of Common Experiments which do not form a single Art.

Histories should also be written of pure mathematics, though they are rather observations than experiments.
129. History of the Natures and Powers of Numbers.
130. History of the Natures and Powers of Figures.

It would not be inappropriate to mention that, as many of the experiments necessarily fall under two or more headings (e.g. the History of Plants and the History of the Art of Gardening will have much in common), the most beneficial thing is to Investigate by Arts, but Classify by Bodies. For we have little concern with the mechanical arts in themselves, but only with those which contribute to constructing Philosophy. But these things will be best decided for each case.

Index

Cambridge texts in the history of philosophy

Titles published in the series thus far

Printed in the United States
53516LVS00005B/1-102

9 780521 564830